HOW TO
LICENSE
your
MILLION DOLLAR
IDEA

*Everything You Need to Know
to Turn a Simple Idea into a
Million Dollar Payday*

SECOND EDITION

HARVEY REESE

John Wiley & Sons, Inc.

For the twins, Jacob and Emma—
the best Million Dollar Idea their parents ever had

Published by John Wiley & Sons, Inc., New York.

Published simultaneously in Canada.

This publication is designed to provide accurate and authoritative information in regard to the subject matter covered. It is sold with the understanding that the publisher is not engaged in rendering professional services. If professional advice or other expert assistance is required, the services of a competent professional person should be sought.

Wiley also publishes its books in a variety of electronic formats. Some content that appears in print may not be available in electronic books.

Library of Congress Cataloging-in-Publication Data:
Reese, Harvey.
    How to license your million dollar idea : everything you need to know to turn a simple idea into a million dollar payday, second edition / Harvey Reese.
        p.   cm.
    Includes index.
    ISBN 0-471-20401-3 (pbk. : alk. paper)
    1. Patent laws and legislation–United States—Popular works. 2. Inventions—United States. I. Title.
    KF3114.6 .R44 2002
    346.7304'86—dc21   2002023467

Printed in the United States of America.

10 9 8 7 6 5 4

PREFACE

There are three types of baseball players—those who make it happen, those who watch it happen and those who wonder what happened.
—Tommy Lasorda

Although the original edition of this book was written eight years ago, hardly a day yet goes by without my hearing from one or more inventors telling me how much they learned from it and how it has changed their lives. I can't tell you how good this makes me feel, and if you are one of the first-edition readers from whom I've heard, I thank you very much for your kind words and for taking the time to express them.

So much has happened since the first edition, to me and to the world, that I should have written this new edition years ago. But I'm happy to have done it now, and I think you'll find it much improved and containing an enormous amount of new material and observations. Of all the events impacting on inventors over these years, none has been as far-reaching as the rise of the Internet. How did we do without it? I've added an entire new chapter on general Internet matters, including web site addresses that I think deserve your attention. Also, specific Internet use recommendations are made throughout the book where appropriate—for instance, how to use the Internet for product research, sourcing help, applying for patents online, obtaining helpful software, seeking potential licensees, and even making appointments.

The second important occurrence is the addition of new Patent Office programs and procedures specifically designed to help private inventors. The most important is the Provisional Patent Application Program, which you'll find thoroughly discussed in Chapter 3. Apropos of this, the Appendix has been totally edited. All outdated patent forms, government office and patent depository locales, source material, inventor club addresses, and sample licensing contracts and agreements have been updated or replaced, and a host of new material has been added.

Finally, after years of saying "no, no, no," a few years ago I decided to offer my services as an agent for other inventors if I thought I'd be successful in licensing their ideas. To announce this and to invite inventors to send me their ideas and inventions, I opened a web site called Money4ideas.com. The response was far greater than I ever imagined it would be, giving me firsthand contact with thousands of inventors around the world. Through this and my free monthly newsletter, I've developed a much clearer understanding of the needs and concerns of new inventors and innovators who would like to put their product ideas to work. In addressing these matters, hardly a page in the book has been left standing without a new insight, observation, or shift in emphasis.

I know in writing this book that I'm not creating an army of competitors but rather a host of new colleagues. Whatever I know, I want you to know. The continuing need for exciting, commercially viable new products is vast and never-ending, and the number of freelance folks to provide them is so small that I can use all the help I can get! When times are good, companies want more and more great new products to participate in the boom, and when times are bad, companies want new products to help keep their ship afloat. Good times or bad, the need for great new product ideas is insatiable. The search never ends. No company ever has too many good products.

I didn't write this book or the first one to be merely a pep talk with little substance. This is a how-to book in the strictest sense, taking you step by step from the very beginning of the process of creating an idea to the very end when the royalty checks start showing up in your mailbox. I hope you find it eye-opening, interesting, and instructive—and I hope it puts you on the path to the kind of lucrative and creatively satisfying new lifestyle that I'm proud to say it has brought to so many others. So what are you waiting for? As Clint Eastwood said in one of his cowboy roles: "Well, are you gonna pull those pistols or just stand there whistlin' Dixie?"

CONTENTS

Success is just a matter of luck.
Ask any failure.

—*Earl Wilson*

The steps in creating and shaping up a simple idea, and how to turn it into a monthly royalty machine.

How to Identify a Market Need and Create the Solution for It
The six-step, can't-miss system for spotting an opportunity, and how to profit from it by creating the perfect product.

Evaluate the Originality and Market Need of your Product and Determine the Need for a Patent Copyright, or Trademark
How to evaluate the merits of your idea, and how to protect it with or without patents, copyrights, or trademarks.

Prospecting, Getting the Appointment, and Preparing the Presentation
How to package your idea professionally, how to select a licensee, and how to get the appointment.

INTRODUCTION

*Money can't buy happiness, but it will
get you a better class of memories.*
—Ronald Reagan

This is a test. What do the following phrases have in common?

It takes money to make money.

The bigger the risk, the bigger the profit.

Go for the burn. No pain, no gain.

The answer: They are all false. Physical therapists tell us that smart exercising needn't cause pain to be beneficial. And I can testify that you don't have to put your savings at risk to turn a handsome profit.

In 1933, during the Great Depression, Charles Darrow, an out-of-work plumber of Allentown, Pennsylvania, found himself with plenty of time on his hands. Just to be doing something, he dreamed up a game about Atlantic City real estate and called it Monopoly. He licensed the rights to Parker Brothers, the big game company, and never had to work another day in his life. That was more than a half century ago, and the royalty checks are still coming in to his heirs. His total investment was some time, ingenuity, and brain power—investment assets that we all possess. If you have the will to put them to use, this book will show you how to enjoy a return on your investment that may potentially exceed your wildest dreams.

SOLID GOLD POUND PUPPIES

The business section of any bookstore is loaded with books that show you how to spend your money. You can become a Wall Street tycoon, you can invest in pork bellies, you can be a real estate mogul, or you can buy a franchise and be a donut shop operator. You name it—just ante up your money and take your best shot. Go for the burn ... and getting burned may be exactly what will happen.

Having a good product idea is only a small part of what it takes to make a business work, and most businesses are gone by the third year, leaving the owner hopelessly in debt. Yes, certainly, yours might be one of the successful ones, but are you willing to bet your house and your children's education on it?

On the other hand, manufacturers pay billions of dollars each year in royalties and licensing fees to people who supply them with profitable product ideas. Incredible fortunes are being made every day by ordinary people who have licensed their valuable inventions and new products to willing takers. Some years ago, a factory worker at a Ford plant near Cincinnati came up with a toy idea called Pound Puppies, which started out as a gift he made for his wife. But he was smart enough to recognize its sales potential and smart enough to realize that he should turn it over to professionals who know what they're doing. So he licensed his idea to what was then the Tonka Company. They sold more than 50 million puppies, and the royalties turned Mike Bowling, the ex-factory worker, into a multimillionaire. Stories like this are commonplace. Ordinary people are reaching for and achieving the biggest jackpots of their lives through licensing. The nice thing is that you can do it, too.

Few ideas are totally original, however, and whatever you dream up, chances are that others have thought of the same thing. The trick is for *you* to be the one to act on it. And later, when the finished product is on the market, let the others be the ones to cry, "Hey! I had that idea years ago!" I call it Slacker's Remorse—or as Sophocles said centuries ago, "Success is not on the side of the faint hearted."

THE HOW-TO-NOT-GO-INTO-BUSINESS-FOR-YOURSELF BOOK

The purpose of this book is to show you how to make your fortune without risking one. If you'll supply the will, I can supply the knowledge. And you can keep your money in the bank. I'm billing this as the *How-to-Not-Go-into-Business-for-Yourself* book. The trick is to have other people spend their money and take the risk. If you have an exciting, new idea that can turn a profit, company presidents all over would love to hear from you, and it would be their pleasure to give you a cut of the proceeds. Any president would be delighted to give you $50.00 for every $1,000 in receipts you can create for his company. It's simply good business.

Licensing is an equal opportunity activity. Good ideas can come from anyone, with or without connections. If it can earn a profit, the manufacturer doesn't care if you're young or old, rich or poor, male or female, black, white, or in between. He doesn't care if the idea comes from Mother Theresa or Attila the Hun. If your idea is a money maker, you will be welcome. To paraphrase a famous quote, except for sheer stupidity, there is nothing that can hold back the impact of a great idea. You can call successful business people lots of names, but "stupid" usually isn't one of them.

TAKE THE MONEY AND RUN

You can earn big money fast through licensing because that's what it is all about. If your idea is really good and you put it in the right hands,

Licensing Is an Equal Opportunity Activity

there's no upper limit to your profit potential. And if you've licensed your idea to a successful, professional marketing company, you'll probably be getting royalty checks while the inventor who wants to do it herself is still looking for funding. For some people, looking for enough capital to start a new business is a lifetime quest. And if they finally do collect the money, the statistical chance of success is very grim. The other way is to turn your idea over to professionals—people who know what they're doing—and then take the money and run. In the movie *Patton*, George C. Scott, who plays General Patton, explains to his troops that war isn't about dying for one's country; it's about getting the other guy to die for *his* country. Well, licensing isn't about investing *your* money. It's about getting the other guy to invest *his.* Let him have the risk and the glory. You'll take the royalty checks.

Perhaps one day, one of your ideas *will* make you a millionaire. But in the meantime, you keep your job, you keep your cash, you keep your house, and you don't get nasty calls from creditors or your bank.

IS YOUR IDEA IN A COMA?

At one time or another, almost everyone has had a clever idea just pop into her head. It just stays there, without nourishment or attention, until it finally slips into a coma. Like Sleeping Beauty, this lovely idea just lies there, day after day, waiting for a Prince Charming to come along. Unfortunately, life's not a fairy tale, and the prince will never show up. What does happen is that eventually this person sees "her idea" on the market and hollers "Foul!" because someone beat her to it.

If you're not prepared to act on your ideas, someone is *always* going to beat you to it. That's what capitalism is all about. Benjamin Disraeli, the nineteenth-century British statesman, said, "Success is the child of audacity." Lots of people get the same idea. The one who does something about it gets the prize, and the others get to complain about having bad luck. Properly exploiting just one useful idea may change your life forever—it happens all the time, and I have a system to help you do it effectively. It's called the C.R.A.S.H. course, and this book is going to carefully lead you through it. You won't be the same when you're finished.

So many readers of this book's first edition have told me how it changed their lives that I have hope and confidence that this new edition will do the same for you.

As Walt Disney said, if you can dream it, you can do it.

CONVERTING YOUR BRAIN INTO AN IDEA FACTORY

Actually, as I'm sure you know, ideas really don't just pop into your head, and they don't come from divine inspiration. You're not going to awaken one morning and say, "Hey, Martha, I just dreamed up a terrific new cure for snoring!" The idea that just seems to appear is actually the end product of a defined step-by-step process. Once you understand what it is, you'll be able to develop ideas almost at will. The process is not automatic; you have to really work at it, but if you follow the prescribed steps, it won't let you down. I create and license new ideas for a living. It's all I do. I can't imagine a nicer occupation—getting paid for my own creativity gives me a rare and wonderful feeling.

THE OPPORTUNITY OF FAILURE

> Don't be afraid to take a big step if one is indicated. You can't jump a chasm in two leaps.
>
> —David Lloyd George

Years ago, I owned a manufacturing company. It was profitable but highly leveraged. I owed my bank almost $2 million. During the Carter presidency, as you may recall, interest rates shot up to an incredible 25 percent. I drained the company's assets trying to keep up with the interest until finally, like thousands of other companies, I was forced to go out of business. More than one hundred people lost their jobs, and I lost everything I owned.

But adversity often leads to opportunity. Mary Pickford, the old-time movie star, once said that failure is not about falling down, it's about staying down. I had to pick myself up in midlife with no money and no job. I took stock of my assets and realized that what I was best at was dreaming

Ideas Don't Just Pop into Your Head

up new products. I could no longer afford to manufacture them, so I decided to turn them over to others for a modest share of the proceeds. I quickly realized that what started out as a desperate course of action was actually a perfect career move.

In my business career, I've met people who have the management skills to run complex, highly successful companies but who have the imagination of a carp. On the other hand, I have creativity to spare, but I had demonstrated to my financial despair that management ability is not one of my strong points. I found instead that management types need me and I need them. It's proved to be a perfect match. I sold my first product idea to the very first company I showed it to. I've never looked back, and I've never been happier. My only regret is the years I spent managing a factory. Many of the companies I originally licensed product ideas to have since gone out of business but new companies come along all the time. They all need new products, and I'm still here to provide them.

WHAT I HAVE THAT YOU CAN HAVE (OR DO HAVE)

I'm not just being modest when I tell you that I'm not a genius—I'm an ordinary guy—and the product ideas I create are not scientific breakthroughs but simple, commercially sensible concepts that a manufacturer can examine and visualize selling profitably. I do have three attributes, however, that separate me from many other people and that have enabled me to earn more money than would have been likely from any other endeavor.

1. I have the confidence that I'll never run out of ideas because I've developed a system for continually producing new ones. It always works, and at any given time I have more ideas than I can handle.

2. I have the discipline to not create or invent until I have a thorough understanding of the industry in which my potential new product is to be distributed. This way I don't invent what's already been invented, and I don't waste time dreaming up solutions to problems that have already been solved.

3. I have the confidence that if my idea is truly sound, I can almost always get a licensing deal for it. I have a battle-scarred, time-tested technique for accomplishing this. There's a system for effectively preparing your idea, getting an appointment to show it to the right people, and saying the right things when you get there. That's the C.R.A.S.H. course, and it's what this book is all about.

There's no mystery to creating and licensing commercially sensible ideas. There are just two easily mastered systems involved: one for developing ideas and the other for getting them licensed. I intend to teach both of them to you in this book because there's more than enough room out there for all of us. Whatever I know, you'll know, and if you apply it, there's no telling how much money you can make.

Research shows that we all possess creative traits. Artists and geniuses don't have a monopoly on creativity, and although you may not be able to compose a sonata, you can be creative in other ways. A simple, workable idea, properly exploited, is also the product of creativity.

Thomas Alva Edison, America's most prolific inventor, always remembered that his ideas had to be practical. He often said that he didn't want to waste his time inventing anything that wouldn't sell. You may not reach his level of genius, but once you get into the swing of it, you'll see that creating commercially viable ideas is really not that difficult. Abraham Lincoln thought up one (a patent for an inflatable system to help boats navigate shoals), and even six-year-old Bobby Patch thought up one (a patent for a toy truck). Just about anyone can do it. The hard part is having the courage and drive to turn it into profit. As has been properly observed, motivation will almost always beat mere talent.

YOU CAN'T MAKE A SCORE
IF YOU DON'T HAVE A GOAL

There are many fine books in print to help you think big, act big, and dress big. But then what? Sure, having a winning attitude is important, but if you don't have a goal to apply it to, what have you accomplished? What good are two highly motivated, all-pro football teams if the field they're playing on doesn't have goal lines? This book offers you a very specific goal to strive for; it shows you exactly how to get there; and, I hope, it will motivate you to get started. If I can help you gain some of that think-big confidence and if I can persuade you to focus it on the goal of creating and licensing commercial ideas, then there's no limit to what you can achieve. Remember: You are what you think, you are what you go for, you are what you do.

CALLING ALL WINNERS

I have books on my desk that tell the stories of inventors, scientists, and researchers who have earned millions of dollars in royalties for their in-

You Can't Make a Score If You Don't Have a Goal

ventions. Certainly these people deserve the credit and riches they've received, but they're not the audience I had in mind when I thought about this book. My original thought was to include some of their stories as positive examples, but I've since changed my mind. It would serve no purpose to explain how Willard Bennett invented the radio frequency mass spectrometer or how Frederick Cottrell invented the electrostatic precipitator or how Gordon Gould invented optically pumped laser amplifiers. I want, instead, to provide profitable information to the person who, in idle moments, may not invent some profound, ultra-hi-tech new system worth many millions but who, quite conceivably, might develop and license simple commercial ideas worth $100,000, $500,000, or yes, maybe even the magic $1,000,000. It happens all the time, probably much more often than you think. Unlimited enthusiasm can bring the most extraordinary results.

If you could meet some of the people I know who have made fortunes for themselves with simple ideas, you'd understand why I say that anyone with reasonable intelligence can do it. Nothing terrible will happen if you fail, but something wonderful may happen if you succeed. The only real losers are those who won't even try. As the baseball guys like to say, you can't steal second if you won't take your foot off first.

I don't want to create the impression that licensing commercial ideas is a simple sure-fire way to make money, because it's not. But what business is? The difference is that participants in licensing prefer to invest instead in their God-given creativity and turn the financial risk over to others. They make the smartest investment of all, in themselves.

Although it's true that almost all of us dream up ideas that have potential value, it's shocking how few of us do anything about it. To some extent, it's because of a lack of knowledge, but to a larger extent it's due to a lack of confidence. Mary Kay Ash, the founder of the hugely successful Mary Kay Cosmetic Company, summed it up well when she said, "If you think you can, you can. And if you think you can't, you're right." Whatever your idea is worth—$10,000, $100,000, or $1,000,000—success is within your reach, and the results may forever improve your quality of life and the well-being of your family. Winners can see an opportunity in every problem. Losers tend to see problems in every opportunity. If you're not already a winner, the information in this book can make you one. If I didn't believe that, I wouldn't have written it.

Free at Last!

FREE AT LAST!

It doesn't matter how much or how little money you have or how far you went in school; if you're a prisoner of a work-for-wages job or have no job at all, this book can set you free. Fortunes are made from exploiting useful ideas. Perhaps it's time for you to make yours. After all, America is still the land where dreams come true.

1

THE C.R.A.S.H. COURSE FOR SUCCESSFUL LICENSING

Ten thousand ideas filled his mind;
but with the clouds they fled
and left no trace behind.

—Thompson

The first real job I ever had was with an advertising agency. One day I had occasion to ride with the president in his car to visit a client. It was many years ago, but I can still remember the ride quite clearly. My boss had a steel bar, about eighteen inches long, somehow welded onto the car's steering column. It extended out past the steering wheel on the right-hand side, and at the end was a little rectangular metal plate with a clip on top. This held a memo pad and a pencil on a string. It looked like what it was, a Rube Goldberg invention, and I shudder to think what it did to the car's resale value. But that didn't seem to matter to my boss. He explained that whenever he bought a new car, the first thing he did was take it into a sheet metal shop to have a contraption like this installed. Because he was in the business of selling ideas, he reasoned that whatever notion popped into his head was a potential company asset. Any idea not captured on paper is an idea waiting to get lost. Ideas are money, and nobody wants to lose money. He understood that until he wrote down the idea, his mind was its prisoner. The mind repeats the idea over and over for fear of forgetting it, and in the meantime all new thoughts are blocked out. It just takes a moment with a pencil and paper to get it back to work.

I've developed this same kind of fanaticism. I trust nothing to memory and immediately write down whatever business thought pops into my head. Random thoughts become lists, lists become outlines, and outlines

11

become plans of action. Outlines give structure to what you do. They unclutter your head and keep your mind on track. They let your brain know you mean business. It's much like an airline pilot's safety check. He may have performed this function hundreds or even thousands of times, but it's too important to do without a written list. There can't be shortcuts—it could mean his life as well as the lives of all his passengers.

I go to a local gym several times each week to use the exercise equipment. I use the same machines in the same order, at the same settings, each time I go. The gym, however, knows the weaknesses of its customers. Every time I go, I receive my personal checklist. After I use each machine, I'm expected to enter the date and check an appropriate block. I'm not finished with my workout until all the blocks are checked. Even here there are no shortcuts. Keeping that checklist up to date keeps me on track in my fitness program, just as other checklists keep me on track in my business.

This book can serve as your own checklist if you do every step and don't look for shortcuts. These systems work and can serve as your guide and mentor. The C.R.A.S.H. program is designed to offer the chance for a degree of wealth, and with it, I hope, an enriched life and the pride that comes from creating something of value for others.

ALL ABOUT THE C.R.A.S.H. PROGRAM AND WHY IT WORKS

There's a passage in *Alice in Wonderland* where Alice comes to a fork in the road and doesn't know what to do. She asks the Cheshire Cat which road to take, and the cat, logically, asks Alice where she wants to go. "I don't much care where," Alice answers. "Then," says the cat, "it doesn't matter which way you walk." You may be reading this book because you, too, are at a fork in the road. Unlike Alice, however, you'll know where you're going, and the C.R.A.S.H. course is going to be your road map. I think you'll like what's waiting for you at the end of your journey.

Gimme a C!

The C in the C.R.A.S.H. course stands for *Create*. Nothing can happen until you create the idea. It's the seed from which can grow a bountiful harvest. There's a simple, proven procedure for sparking ideas that works for me and for just about every other creative person I know or have read about. I'm confident it will work for you as well. And even if you already have plenty of good ideas, I think you'll still find the suggestions in Chapter 2 interesting. Nobody has ever suffered from too many good ideas.

Gimme an R!

The R in the C.R.A.S.H. course stands for *Research.* Is it really a good idea? Is it original? Can it be done? Will it sell? Assuming you come up with the right answers to these questions, you need to know how you can protect your idea in a safe, prudent, and practical manner.

I spend a great deal of time in the book on this step. Whenever I lecture to aspiring inventors, the questions I'm most frequently asked center on protecting an idea. I've met people who are so frozen with fear that someone is going to steal their ideas that they do absolutely nothing with them. Patent attorneys tell me they get calls from inventors who inquire about fees, but won't reveal their ideas, even in the most general terms. One day, these inventors will be carried off to the old folks' home with their precious ideas still locked within them. Did they have ideas worth millions? Could they have lived out their lives in splendor? We'll never know and neither will they ... but their ideas are safe.

In Chapter 3, I will show you simple, inexpensive ways to protect your concept, often without the need for an attorney. But if you do need one, I'll show you how to keep fees at a minimum. Nobody hates to pay legal fees more than I do, and it's surprising how much you can accomplish on your own.

Gimme an A!

The A in the C.R.A.S.H. course stands for *Action.* You have the idea. It's innovative, it's protected, and now's the time to show it to the world. This step, discussed in Chapter 4, deals with the preparation of your presentation, the selection of prospects, and the tricks to get a face-to-face appointment with the decision makers. There are professional effective ways to do these things, and I'll explain them to you.

Gimme an S!

The S stands for *Show 'n' Tell.* This is make-or-break time. It's when you go out and sell your ideas to prospective licensees. In Chapter 5, we'll discuss what to show them and what to tell them. I'll give you tips on how to make presentations and prep you on what questions to expect and how to answer them. This is also the right time to discuss the potential use of agents. Do you need them? How do you find them? How do you negotiate with them?

Gimme an H!

If the creation of an idea is a seed, the other steps in this program are designed to care for and nurture it until maturity. Now it's time to reap the *Harvest.* And that's the H in C.R.A.S.H. Exactly how bountiful the harvest will be, and in what form and under what circumstances, must be spelled out in a contract. This is the payoff time, and Chapter 6 will help you get everything you're entitled to.

A sample of the contract I use myself is included. If you follow it, you probably won't need an attorney; if you do use an attorney, you'll be a better client for understanding the process. I'll point out what provisions must be included for your protection, what 10 provisions are negotiable (and how far), and what 10 provisions you must never agree to. You will have worked hard to get to this stage, so you don't want to blow it with an unsatisfactory agreement. Follow this chapter carefully and that won't happen.

I can't think of a better way of summing up the C.R.A.S.H. course than by quoting the first-century Roman philosopher Epictetus:

> No great thing is created suddenly, anymore than a bunch of grapes or a fig. If you tell me you desire a fig, I must first answer that there must be time. Let it first blossom, then bear fruit, then ripen.

What was observed almost two thousand years ago is no less true today. The C.R.A.S.H. course is designed to guide you every step of the way from seed to harvest. And now we start. It's planting time. I hope your crop is bountiful.

By the way, someone who knows how to nurture the seed of creativity must have seen my old boss's car, because soon afterwards a terrific product appeared on the market that's a big improvement on his old welded steel bar. I'm sure you've seen it. It's a notepad clipped to a plastic base that fastens to the dashboard by a suction cup. It's been around now for years, and every automotive store sells them. I wonder if my old boss complains that someone stole his million-dollar idea? It was staring him in the face all those years, and he never realized it. If he had taken his idea and licensed it, he would probably have made more money than he ever did with his advertising agency.

2

CREATING THE IDEA

How to Identify a Market Need and Create the Solution for It

I don't care a damn for the invention.
The dimes are what I'm after.
— *Isaac Singer*

There is a widely held misconception that creativity flourishes best in an unstructured environment. However, interviews with creative people show that their environments and work habits tend to be quite regimented. This self-imposed discipline allows them to get the work done. People who rely on the creation of original ideas as a profession have long understood that it is not a haphazard activity. Otherwise, how could they stay in business? The work structure functions like the banks of a river. Without them, the river meanders all over until it simply disappears. Creative thought requires similar boundaries in order to be effective.

A clearly defined system is at work in the creative process, even if the participants follow it subconsciously. Once you recognize what's going on and break it down into steps, you can condition your brain to produce ideas almost at will. Only six steps are involved, and they are certainly not complicated.

Libraries are filled with books of theories about the creative process. The subject has fascinated investigators for centuries, and in many respects it's still a mystery. The consensus, however, is that although few of us are geniuses, virtually all of us have far more creative ability than we could possibly imagine. Apparently no direct correlation exists between an exceptionally high IQ and creativity, which has more to do with your mind-set and the amount of discipline you bring to the task. If you look for ways to be creative—if you really work at it in a systematic way—the results may astonish you.

As a mnemonic device to explain my creative process, I took the first letter of the action word in each step and put them together. They spell *icicle*. You would think I was terribly clever if I could get them to spell something like *monies* or *create*, but unfortunately they don't. All they spell is *icicle*.

To follow Reese's I.C.I.C.L.E. system for creating commercial, profitable ideas, you must

I. *Identify your general goal or objective.* Define the general problem.

C. *Concentrate on developing a solution.* Do research.

I. *Identify your goal again.* This time, pinpoint it.

C. *Concentrate again.* This time, focus on the narrowly defined goal.

L. *Let it go.* Go to sleep; go to a party. Let your subconscious go to work.

E. *Eureka!* When you least expect it, the idea just pops into your head.

I realize this sounds too simple to be true, but if you'll give I.C.I.C.L.E. a chance, you'll be astonished at how creative you can be. Most great ideas are characterized by their simplicity, and an idea about creativity itself needn't be an exception. If you read through the books on the psychology of creativity, you'll see that although researchers remain perplexed about *why* thoughts and ideas are created, they've long understood *how*. This is the classic evolution of an idea, a thinking process in use since people began creating useful objects. Psychologists may have more learned ways of describing it, but it's still just Reese's I.C.I.C.L.E. program.

Before moving on to the I.C.I.C.L.E. program itself, I'd like to make an important point. Having seen literally thousands of aspiring inventors' new product ideas, one overriding problem is evident. Most inventors invent first without having a true understanding of the marketplace in general— or in particular, of the specific industry in which their product idea is intended to compete. The resulting idea is often simply naive, the solution to a long-ago-solved problem, or an idea with scant commercial value. It doesn't mean these folks aren't capable of coming up with something wonderful, but their chances are greatly diminished by working in the dark as they do. Success for them is often the result of plain dumb luck.

It's sort of like a tailor who makes a suit and miraculously finds someone it fits and who happens to need just what the tailor sewed. Wouldn't it be best to first locate the customer, determine her needs, take her measurements, and then make the suit?

The solution is simple. Remember that inventing is a business—and that your customer isn't the consumer. Your customer is the guy who *sells* to the consumer. If you acquire an understanding of the environment in which he functions, your chances of giving him what he's looking for are greatly improved. A hunter who understands his prey is always more successful than some guy who just goes clomping through the woods.

STEP 1. *IDENTIFY YOUR GENERAL GOAL OR OBJECTIVE*

Now, assuming you have a working knowledge of marketing in general and your industry in particular, you're ready to get started. The creative process begins the moment you pick up a pencil and commit a general goal or objective to paper. A daydream ends; a plan of action begins. Just as you can't get your troops to march if you don't give them marching orders, so you can't get your brain into production until you've told it what to produce. As the old saying goes, If you don't know the harbor you're headed for, no wind is the right wind. Defining a goal, or selecting a harbor, sets your imagination on an exciting voyage with a crisp wind filling its sails.

**Creativity Starts
When You Reach for a Pencil**

The most difficult part of the idea-making process is providing a clearly identified and stated goal. A clear definition often suggests the solution. The trick is to recognize it. If, about 10 years ago, I had said, "Do you notice the tough time people have carrying their luggage through airports?" It wouldn't have taken you long to create those collapsible little carts that were sold by the millions. And then, a few years later, someone said to herself, "Why carts? Why not simply put wheels and a handle on the bag itself?" Any one of us could have designed the carts or the bags, but the one who got rich was the person who had the imagination to recognize the problem. Jacques Lipchitz, the famous artist, once defined creative imagination as the ability to invent the wheel while observing a man walking.

Plunging Ahead

A number of years ago, I noticed this headline in *The New York Times* (June 13, 1992, sec. 3, p. 17):

TOILET PLUNGER IS THE MODEL FOR DEVICE TO RESTART HEARTS

How can you resist reading a story like this? Here's the first sentence:

> A mother and son in San Francisco, who used a toilet plunger to revive a man whose heart had stopped, have inspired development of a simple new suction device that some experts say works better than traditional cardiopulmonary resuscitation.

The article proceeds to tell how the new device works and that it could help save some of the more than five hundred thousand Americans who die of heart problems outside a hospital each year. The boy who saved his father with the toilet plunger (he did it twice over several years) suggested to doctors that they have one by the bed of every cardiac patient. According to the article, most of the doctors simply found this to be an amusing suggestion and went about their business. Opportunity was knocking, but the physicians weren't home to callers.

Three doctors did take the suggestion seriously, however, and after design work and testing, the Cardio-Pump was born. Patents have been applied for, and a manufacturer has been licensed to manufacture and distribute the product worldwide. If it works as well as it is supposed to, you can assume the doctors who hold the patent will never have to lift another tongue depressor for the rest of their lives. And do you suppose the other doctors are still amused?

Being Part of the Solution

Problems like this are all around us, just waiting to be solved. Even so obvious a concept as a left shoe and a right shoe was only conceived a little more than one hundred years ago. Most of us are not part of any problem, and we're not part of any solution. We're just sort of standing around not paying much attention to either. But if you can get yourself to become part of the solution, you have a good chance of becoming well off—and you may even become rich.

The starting point is to develop the belief that you can do it ... and that you *deserve* the rewards of your accomplishment. Losers are not only convinced that they can't do it but also think it's part of some grand universal plan that they should be where they are in life. It's as if, to them, their personal success would spoil some cosmic system. And then who knows what would happen? It's the *qué será, será* syndrome.

Winners even do better in hospitals. Studies have shown that people who fight for their recovery by calling the nurse all the time, demanding to see the doctor, always asking for medicine, and in general just being a demanding patient have a much better survival rate than those who lie in their bed quietly and say, "Whatever will be, will be."

Turning Irritations into Opportunities

To uncover a goal or objective, the logical place to start is with what you know: home, hobbies, work. Just as pearls come from irritated oysters, product ideas often come from common, daily annoyances. Is there some chore at home you hate to do? Would some kind of new product make it easier? Whatever irritates you is most likely irritating millions of others as well. The person who first came up with the idea of a little frame to hold a leaf bag open was undoubtedly less than delighted when raking the lawn. I bought one of these gadgets a few years ago, and I bless the inventor every time I rake my leaves. It's hardly a breathtaking product and shouldn't even be mentioned in the same breath as the Cardio-Pump, but it answers a need and provides a profit. What more is necessary?

Climbing the Corporate Ladder

Many new ideas are work related for obvious reasons. You see a problem or bottleneck on the job, and your creative mind starts to think of a solution. To avoid problems later, and perhaps to get the rewards you're entitled to, it's a good idea to find out your company's policy on new product development. If you're entitled to be compensated, you should know it up front.

Proper ownership of an idea is easily recognizable at both ends, but the middle ground can get murky. If, say, you're a computer program operator with a medical company and, on your own time, you develop a new type of fishing rod, this is clearly none of your employer's business. At the other extreme, if you're a scientist working in the company's laboratories and you develop an exciting new chemical process, this obviously belongs to the company. But suppose you work in the shipping department and, borrowing some of the laboratory's materials and asking a company chemist for some advice, you work at home on your own time and develop a new method for coating pills. Whom does this belong to?

The idea for Band-Aids was developed by a then low-level employee at Johnson & Johnson, and he automatically turned it over to the company. Is it any wonder that he retired years later as a senior vice president? Also, the scientist who developed the Post-It Notes is a 3M employee hired to invent, and the product obviously reverted to the company. Large companies all have clear policies for regarding employees in such situations. If your company is small and doesn't have a well-defined policy, I don't think it's out of line to ask that you be given a memo of understanding before you submit any ideas to them. Once they already have your idea, it's a little too late.

When Is an Idea Not an Idea?

I've come to realize that when most people complain about "their idea" being marketed by someone else, they never actually had an idea to begin with. What they really had was an idea *for* an idea. They didn't actually sit down and work out the design for a product to hold a plastic leaf bag open. They didn't apply for a patent, they didn't make a prototype, they didn't do any research, and they didn't develop a marketing plan. All they did, a few leaf-raking seasons ago, was say, "Dammit! Somebody oughta come up with a contraption to keep these damn bags open!"

Well, their wish was eventually granted, and somebody did. Nevertheless, when the contraption finally did appear on the market, thousands grumbled that they had the idea first. But they didn't have an idea; they had a thought. You can't license a thought, but you can profit mightily from a fully conceived, designed, and developed idea.

After years of saying "No! No! No!" I recently decided to offer my services as an agent for other inventors whose products I like. I opened a web site for this purpose, Money4ideas.com, and the number of submissions has been overwhelming. I'm amazed at how wonderful some of the ideas are, and I get excited about getting great licensing deals for these folks. But alas, the bulk of the submissions reveals the sad fact that many inventors simply have almost no knowledge of the nature of a licensable idea itself.

First of all, an inventor must understand what the licensee is paying for. The licensee, usually a manufacturer, gets up every morning thinking about ways to get ahead of her lousy competitor across town or across the country. She's constantly trying to exploit the other guy's weaknesses while promoting her own strengths. It's a very serious battle, and it never stops. Simply stated, if you come along with some kind of weapon that will give one of these warriors an edge over the other, she'll gladly pay you for it. What she's *not* going to pay you for are mere suggestions, simple product adjustment, tweaks, or ideas that she has to do the work on. She wants you to give her a new sword for her battle, not just a pound of steel and a rough pencil sketch.

Here's an example of what I mean: "Hey, Harvey, I have this great idea for a new toy. It's a teddy bear, with green fur, and when you squeeze its paw, he jumps high in the air, does a double flip, lands on his behind, opens his mouth and says, 'YO, MAMA!' "

"Wow, that sounds great!" I say. "How does it work?" The inventor looks at me funny and says, "I already told you—there's a motor inside and you squeeze his paw to start it."

Here's another example. There's a certain product that comes in 6- and 12-ounce sizes, and the "inventor" determined that sales would greatly increase if the manufacturer also made it available in a 4-ounce size. It made perfect sense to me—until the inventor asked me to get her a royalty deal on each 4-ounce unit sold.

Many of the product submissions I receive from outside inventors are informal little suggestions, ill-conceived ideas that are at best fodder for a helpful hints column. It makes me wonder what people's perception of inventing and licensing is really all about. I don't expect an inventor to know how to make a dramatic visual presentation, how to get to the right licensee, or how to negotiate a contract. That's why there are people like me. However, I *do* expect to see something of value, a fully formed idea based on some experience or knowledge. My company will dress up the design, make the prototypes, draw the pretty pictures, make the mock-up packaging, and knock down the doors. However, the inventor has to at least supply us with something real, something that works and that gives the licensee enough of a competitive edge so that we can get his signature on the contract.

Get It in Writing or Kiss It Goodbye

Another approach is to watch the people around you: at work, in the malls, on the streets. See what they do, listen to what they say, observe how they're dressed. I do a lot of mall watching and have developed some very good ideas in this fashion. In the immortal words of Yogi Berra, "You can observe a lot just by watching." Whatever you do, always carry a notebook and write down any random thoughts that come into your head.

My permanent idea book contains several years' worth of observations, thoughts, and rough doodles. When I review it, I'm able to see a variety of new product ideas beginning to take shape. I try to prioritize these ideas and turn my energies toward developing what seem to be the most promising. At the same time, as I discard passé ideas, I am entertaining new ones. It's an ongoing process, and my inventory of new ideas is always larger than I can develop and sell.

You Do It because You Do It

It's my discipline that I never go anywhere without a notebook and pencil in my pocket. They're on my dresser, next to my wallet, and automatically go into my pocket each morning along with my wallet and my money. Musing over a new product idea is never far from my thoughts, and if I think of something, I can't rest until I've captured it on paper. I never doubt that I'll come up with my next new product.

I sometimes wonder how great cartoonists can come up with such funny cartoons every day, week in and week out. My guess is that like me, they're never without a notebook and pencil, and thinking about new cartoon ideas is never far from their minds. It's what they have to do, and they just do it. I'm sure they never doubt that they'll come up with their next cartoons. Not ever.

If you tell yourself that it's what you have to do, you'll do it, too. Belief releases the creative juices, so never doubt that you can do it. Just make a start and the rest will follow.

Send an E-mail to Your Brain

Your brain registers untold thousands of sights, events, and impressions every day. There's not a waking moment when some occurrence isn't being absorbed. It's a deluge, and your poor brain is so busy just taking notice that it sometimes has difficulty separating the important from the humdrum. That's where note taking comes in. It's the way to tell your brain, "Hey, this is really important, so file it somewhere that's easy to reach." Don't try to evaluate your thought on the spot; instead, write it down and worry about it later. Afterward, when you've had time to reflect, you can review your notes and separate the valuable from the valueless. Before you know it, by periodically reviewing your notes, a course of action will become clear and you'll be on your way. So buy a notebook. It's the best 49¢ investment you'll ever make.

Finding Your Niche

I happen to create many toy products, so one of my methods for establishing goals is to wander through toy stores looking for niches and gaps to fill. This is what corporations do all the time. One of the world's largest giftware companies, for instance, has trained people who continually visit stores, malls, and specific events, hoping to spot trends and identify new product opportunities. They call it *niche marketing*, and I figure my job is to beat them to it. If I'm interested in developing a new toy, I try to put myself into the shoes of a toy manufacturer and determine what someone in that business would like to produce. I almost always come home from the stores with ideas and rough thoughts. I do the same thing in the sporting goods industry, the home furnishings industry, or any other one I am targeting at the moment. For instance, on page 24 you'll find a letter I recently received along with a catalog from one of the leading toy companies.

To proceed with our program, let's assume that your hobby is camping, that several camping manufacturers are in your area, and that you've decided to look for a new product for that industry. This is all you require

Harvey Reese Associates

Att: Mr. Harvey Reese

Dear Mr. Reese:

    Thank you for coming to visit us at Toy Fair. As you witnessed first hand, good things are happening at Playtime. We have many new product categories and fresh new products throughout our line <u>including seventy individual products from outside inventors/designers</u>. How many companies can claim a number like that?

    In   we will be adding to all areas of our existing business and will be <u>expanding into new categories like sports and outdoor products</u>. <u>We are looking for innovative new additions</u> to our existing lines and new lines or categories of products to expand our business.

    Attached is a Playtime and a Helm catalog for your reference. We're counting on you to show or send us more exciting products this year!

    I look forward to seeing or hearing from you soon.

                              Best Regards,

                              PLAYTIME PRODUCTS, INC.

                              Loren Taylor
                              Loren Taylor
                              Executive Vice President

LT/js

Enc.

Reprint of Playtime Letter

for the first step. You have your general goal. You've sounded reveille for your brain.

STEP 2. *CONCENTRATE ON DEVELOPING SOLUTIONS*

In Step 1, you identified a simply stated goal or problem to solve: to create a new product for the camping industry. In Step 2, you must concentrate on the camping industry and find out everything you can about it. You're look-ing for a niche, and that requires putting the industry under a microscope. Go to libraries, go to stores, talk to people, read magazines, and write down everything you see and learn. Fritz Maytag, former president of Anchor Hocking, writing in the *Harvard Business Review,* said, "If you're going to make rubber tires, you better go to Malaysia and see the damn rubber trees!"

By concentrating on the camping industry, you'll be looking to uncover the specific need for a new product or important ways to improve existing products. In Step 1, you sounded reveille for your brain. Here, in Step 2, you prepared to give it its marching orders. You're seeking a target of op-portunity, and when you've found it, it's time to move to Step 3.

STEP 3. *IDENTIFY YOUR GOAL AGAIN*

Your originally stated goal needn't be anything more than a rambling, loosely stated direction in which to proceed. "I want to do something in the camping field" is good enough. If you live in Baltimore, Maryland, for instance, and you want to drive to your friend's house at 123 Maple Street in Denver, Colorado, your general goal is simply to head west and find your way to Denver. When you arrive there, you focus on your narrower goal of finding 123 Maple Street.

For the sake of discussion, let's say your research uncovered a need for a new type of disposable, biodegradable eating utensil. That's the niche you were looking for, and you're now able to instruct your brain, with preci-sion, where you want it to concentrate. You restate your goal in a precise way by saying to yourself, "Forget about blankets, tents, boots, and cots; my objective is to develop a completely new line of *biodegradable eating utensils that can be reused or disposed of,* whichever the camper wishes." You have a 10-word objective, so it's time to move on.

If Step 1 was your brain's reveille and if in Step 2 you gave out marching orders, Step 3 is where your brain starts marching toward its destination.

STEP 4. *CONCENTRATE AGAIN*

When sportswriters talk about a star athlete, they usually say something like, "He has great hands and terrific concentration" or maybe "She has all the natural skills and terrific concentration" or often "He has enormous speed and terrific concentration." Can we detect a pattern here? Having this unique ability to give 100 percent of their attention to what they're doing seems to be the one trait that all great athletes have in common. And that's what's required of you here—not casual concentration but *intense* concentration.

Now that you have a precisely defined goal, you have to bring all your focus to bear on it. You should be thinking about camping utensils day and night—making drawings, trying experiments, making prototypes, filling page after page with sketches and doodles, doing research, reviewing your material over and over again—and all the time thinking, thinking, thinking.

Just as you spend quality time with your children, so you have to spend quality time with your idea. And quality time here means concentrated time. No radio, no television, no telephone. Just you and the problem, eyeball to eyeball. Personally, I like to take long, solitary walks. It's been said that angels whisper to a person when he goes out for a walk.

How to Find a Fresh Idea

The creative process at its most primary level means thinking about a problem. It stays on your mind and invades your soul. The solution is so very close, but it disappears in a puff when you look in its direction. Tons of books describe methods to help you think creatively. They advocate techniques such as creative environments, quiet spaces, calmness, free association, and inner freedom. They tell you to do everything from solving children's puzzles to sitting in a tub or standing on your head. They show you special ways to sit and secret ways to breathe and mysterious ways to roll your eyes, all in the name of creativity. However, although sometimes the methodology may be laughable, the objectives are quite serious.

In a few simple words, all the juvenile puzzles and body-stretching techniques represent efforts to get you to think about a problem in an open, unfettered manner. Certainly I support that objective. If you're constricted in your thinking by convention, it will stand to reason that your thinking is going to be conventional.

I know a great many creative people, and no two of them work on their creativity in exactly the same manner. If you do it best by standing on your head, who am I to complain? What every creative person is trying to achieve is a deep level of concentration. How you accomplish it is up to you. Three famous authors being interviewed about their work habits will show very little similarity in their answers. One may tell you that she writes longhand in her kitchen from 12 midnight until 5 A.M. Another will say that he has a computer in his den that he works on every day for as long as it takes him to write seven pages. The third writer may reveal that she dictates to a secretary every morning while still in bed. They didn't get these techniques from a book, and they'd probably never suggest that budding writers follow these work habits. They merely found what works best for them, as I'm sure you'll find what works best for you.

Making Boredom Pay Off

If you ever exercise on a treadmill, then you might agree it's the longest, dullest 30 minutes you could ever spend. My health club understands this, so it locates most of the treadmills up front where there are lots of TV sets, lots of mirrors, and lots of comings and goings by the members. In the back corner of the gym, a few more treadmills are in operation—no television programs, no distractions, relative peace and quiet. These are the machines I use, and I find it to be the most creative half hour of my day. There's no place to go and nothing to do, so my mind can gnaw on a problem like a bull terrier with a bone. I'm not alone. Men and women on either side of me are lost in their own problem solving. None of us got this out of a book on how to be creative; it's just some thing we independently discovered. Thinking time is where you find it. Surveys show that many people get their best ideas while showering or on the toilet. So keep a notebook handy because you never know.

Learning from Leonardo

Treadmills and toilets aside, over the long haul what works best for me and for most other people is simple doodling. Nothing surpasses it for locking your attention onto the problem. A pencil must be the best concentration device that's ever been created. It's the tool people have used to solve problems for many centuries. In his diaries, Leonardo DaVinci, perhaps the most creative person in history, noted that his ideas came while doodling (he called it scribbling). What was good enough for Leonardo certainly should be good enough for the rest of us.

All research into the fascinating study of creativity indicates that results come not so much because of the creative technique itself but from the level of intensity applied to a problem. Someone once asked a famous scientist what made him so special. He said it was because he could think about a problem for 15 minutes straight and other people couldn't. It sounds simple, but it's not. Try it sometime. A person who can focus on a problem that intently *is* special! But even if you can't match that performance, it's important to give it your very best shot.

The Assumption of Success

The other key factor in the creative process is what I call the assumption of success. By assuming you'll find the solution, you significantly improve your chances of doing so. This is where Dale Carnegie's power of positive thinking comes into play. Although I can't explain why it works, developing a winning attitude has helped thousands. We can only become what we first behold in our imagination.

Athletic coaches, regardless of the sport, all teach the assumption of success. If you're a golfer lining up a putt, you visualize the ball's route from the putter to the hole. If you're kicking a point after touchdown, you visualize the ball going through the uprights. You're seeing in advance the success of your endeavor—you simply can't succeed in sports unless you're able to do that. Calling it a winning attitude is not quite accurate. It's a winning *assumption*. No successful major league batter ever steps up to the plate without expecting to get a hit. Life's endeavors work the same way. Picture your creative work to be successful, and your brain won't let you down. It really works.

People who create—designers, writers, inventors—all expect to be successful in whatever they happen to be working on at the moment. An author doesn't start writing a book without expecting it to be a good one, and an inventor doesn't embark on a project without expecting to bring it to a satisfactory conclusion. If they didn't have this assumption of success, they'd have to look for different careers.

So if you've concentrated on the problem for as long and as hard as you can and if you've developed the winning assumption about the outcome and if you're brain weary from the effort, it's time to give it a rest.

STEP 5. LET IT GO

It's now time to slam your notebook shut and go to the movies, go to bed, or go to a party. Forget about the problem and give your brain a break. This is the incubation period. Your brain, now on automatic pilot, has an incredible capacity to sort through all the bits and scraps of information you've acquired about the problem. Scientists have long recognized this phenomenon, and yet how it works remains a mystery. That it does work, however, is undeniable. Your subconscious sorts it over, mulls it over, and when you least expect it, the most wondrous thing happens....

STEP 6. EUREKA!

You know what it's like when you can't remember someone's name? It's on the tip of your tongue, but you can't get it out no matter what you do.

It's Time to Give Your Brain a Rest

You think and think as hard as you can, but it just won't come. Later, seemingly out of the blue, perhaps while you're teaching limericks to your parrot, you suddenly snap your fingers and say, "Herman Smedley!" All of a sudden, you can remember everything you ever knew about dear old Smedley, down to the mole on the back of his neck.

The information was always there, of course, inside your brain. Your subconscious merely took its own good time to process it and bring it forward. That's exactly how good commercial ideas often are formed. Have a clearly defined goal, store the appropriate information, and give your brain the time to process it. Either the idea will pop out fully formed or, more frequently, a clue will emerge to put you on the right path. Either way you're off and running.

Unfortunately, as you've probably already guessed, Reese's I.C.I.C.L.E. process is not without its flaws. You sometimes have to repeat steps; it will only solve problems that are within the brain's capabilities; and, most important, it takes its own sweet time in doing it. The process may take seconds, days, weeks, maybe even years. And there's an element of luck and happenstance involved that can't be controlled.

And yet it works. I can remember a recent event when luck was on my side and I was able to go from a dead stop to identifying the problem, creating the solution, and actually selling the new product concept all in one day.

AN UMBRELLA FOR A FISH

I was wandering through a toy store, looking for some direction, when it occurred to me that I should focus on developing a new summer toy *(identify your goal)*. I reasoned that because fewer toys are available during that time of year, a toy manufacturer would welcome something to balance against the Christmas season. I wandered over to the display of summer toys *(do research)*. What quickly became obvious was that the best summer toys all involved water. I was now able to narrow down my objective *(identify your goal again)* to creating a new toy that would involve water. I reached this stage in well under 15 minutes.

I thought hard about the problem *(concentrate again on the narrow goal)* while driving to keep a luncheon date and then gave my brain a rest *(let it go)* as I met my luncheon companion. Later, while driving home, the germ of an idea began to form *(Eureka!)*, and I couldn't wait to put it on paper to see how it could work. In short order, I had the bugs worked out and faxed the concept off to a company with whom I do quite a bit of business. Within an hour I received a telephone call advising that everyone at the company loved the idea and that we had a deal and I should send them a

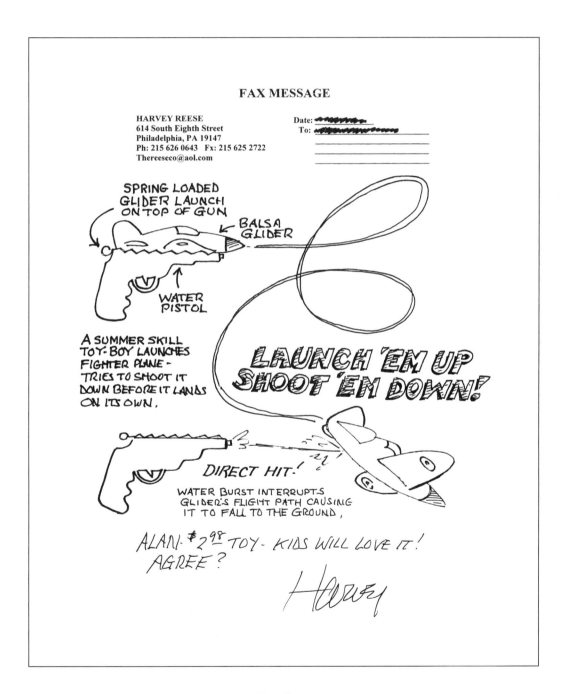

Fax Copy

contract. Rather than describe the item, I've reproduced the fax I sent on page 31.

It's not rocket science, but if all goes well, lots of kids will be happy, lots of shopkeepers will be happy, and I'll be happy. I'll grant you, this product is about as useful as an umbrella for a fish, but what has rocket science done for us lately?

I'm often asked if I'm fearful that someday I'll run out of ideas. I reply honestly that I'm not. As simple as it appears to be, the I.C.I.C.L.E. system works for me, just as it has for others since the first time a human made a conscious effort to be creative. A system doesn't have to be complex to be valuable. As long as I can identify a problem and as long as I do my homework, I'm confident the system will provide the solution.

Einstein was once asked his system for getting good ideas. "Actually," he replied, "I've only had one good idea, and that's just a theory." Perhaps one good idea is all you'll need. If you exploit it properly, it may change and enrich your life. As they say, getting an idea should be like sitting on a pin: It should make you jump up and do something. In the next chapter, I'll discuss what that something should be.

3

RESEARCH

Evaluate the Originality and Market Need of Your Product and Determine the Need for a Patent, Copyright, or Trademark

The harder you work, the luckier you get.
—Gary Player

Now that you have a commercially viable idea, it's time to research and protect it. In this chapter, we'll discuss the novelty of your idea and whether it merits the expense of a patent application. Then, we'll talk about how to go about obtaining a patent.

The U.S. Department of Commerce publishes an informative booklet called *General Information Concerning Patents*, which you can buy for a few dollars. Refer to the Appendix for ordering information. According to this booklet, "Any person who invents or discovers any new and useful process, machine, manufacture or composition of matter, or any new and useful improvement thereof, may obtain a patent." This quote refers to a utility patent, the acquisition of which is a long, very expensive process—and often, to the chagrin of many inventors, totally unnecessary.

Many perfectly fine new product ideas either don't qualify for a utility patent or don't warrant the expense. That doesn't necessarily make these ideas any less valuable. You can probably still get ample, appropriate protection at a modest cost. There are design patents, trademarks, copyrights, and a whole host of other ways to protect your idea, and even if your idea doesn't qualify for any of the traditional methods of legal protection, it can still be quite valuable. Alfred Hitchcock used to say that if didn't matter if the hero in his movies was looking for a military code, stolen jewelry, or a secret formula. He called that object the *McGuffin,* and its only purpose was to move the action along. The adventure itself (with a happy ending) was the real story.

In our movie, the quest for profit is the real story, and a new product idea is just the McGuffin. People aren't interested in your invention as such or your merchandising idea or your book or your play. They're interested in how it can be used to create profit. Any new and original idea that can create profit by its use has value and can be licensed—whether it's patented or not may be immaterial. Here's an example:

A number of years back, when soap operas were at the height of their popularity, it seemed as if everyone in the country was watching *General Hospital.* College kids were cutting classes, and people were actually staying home from work to watch "important" episodes.

Thelma Reese, a college professor, and some friends, spent a riotous lunch one day trying to connect the relationships between the various characters: who was married to whom; who was divorced from whom; who were really brother and sister but didn't know it; who were the real fathers of which kids; who was switched at birth with another baby in the hospital; who was really a murderer; who had amnesia and was not really the person he thought he was; who was plotting murder; and who was plotting the secret overthrow of the government. And that's not the half of it.

For the others it was just a funny lunchtime exercise, but for Thelma it was the germ of an idea. A few weeks later, after fully developing the concept, Thelma met with executives from ABC Television in New York and convinced them to give her the right to publish a full-color poster to be called "The Official History of *General Hospital.*" It would be part history, part storyline, and totally funny.

Thelma then made a deal with a poster company that agreed to give her 40 cents for each poster sold. The company sold 110,000 copies, and Thelma became the Dr. Joyce Brothers of the soap opera world. Lecture agents wanted to put her on tour. The editor of a soap opera magazine called to discuss the possibility of a monthly column. Although she turned down most of the overtures, it was a glorious 15 minutes of fame.

This story illustrates that you don't have to be an inventor with a patented product or a secret formula to earn royalties. Thelma couldn't go to the Patent and Trademark Office (PTO) with the idea to make a history of a soap opera, but she made it a fully developed concept by securing the exclusive permission from ABC Television. And then she found a company that saw money-making potential in her McGuffin. Thelma earned $63,000 in royalties. She also earned some free trips to Hollywood, appearances on network and local TV shows, and guest appearances around the country on radio talk shows. She was the subject of many newspaper articles and was offered the possibility of a lecture tour (which she turned down) and the possibility of a monthly column (also turned down).

The royalty money, not a princely sum to begin with, went quickly. Fame, as they say, is fleeting. What did remain was a new sense of pride and self-confidence that has enabled her to achieve more professionally than she ever dreamed possible, including recognition in the form of invitations to White House luncheons. New ideas fly off her like sparks off a pinwheel, and some have turned out to be quite significant.

This true story highlights several facts:

1. The absence of a patent is not the death knell for a product idea. Often, it simply doesn't matter.

2. You can't merely have an idea for an idea. You have to spend the time to create a fully developed product. The more you leave for the licensee to do, the less valuable your idea becomes. If Thelma had gone to the poster company and asked for royalty money simply because she had the notion that a *General Hospital* poster would sell, she wouldn't have gotten to first base. Instead, she offered the idea, the ABC exclusive, research about the sales potential, a demographic profile, and a fully developed sketch showing what the poster would look like. What she did, in effect, was make them an offer they couldn't refuse.

3. Finally, this story illustrates that *any* commercially viable product can be licensed if you find the right buyer. The "new product idea" to which I keep referring needn't be a three-dimensional item that you can hold in your hand. Thelma's product was simply a turnkey scheme to make money. Let your imagination be guided by the profit, not the patent. Always remember that you're not in the invention business; you're in the McGuffin business.

Larry Homes, the great boxer, has summed up the concept much better than I ever could: "Why do you think I'm fighting? The glory? The agony of defeat? You show me a man who ain't fighting for money, I'll show you a fool."

IS YOUR IDEA REALLY ORIGINAL? SO HOW COME NOBODY THOUGHT OF IT BEFORE?

Probably a dozen or maybe a hundred people have thought of your exact idea before you did. That's beside the point. What's important is that none of them ever did anything about it. If you do, then it becomes your idea to keep. If you *don't,* then you're just part of the mob. I make money on some of the most trivial, obvious ideas you can imagine. It's incredible that nobody did anything about them before; it defies logic. And yet, of all the presentations I've made over the years, in only one instance had the company previously thought of that identical idea. Even then, I licensed the idea to one of the firm's competitors and collected healthy royalty checks on it for several years thereafter.

Don't Reinvent the Wheel—Someone Else Has Already

The reason I'm unconcerned about someone else having the same idea is because I thoroughly research it before moving forward. If a similar product presently exists or if it's an old idea that existed before, this is the time to find out. Your research may be as simple as visiting a few stores, talking to a few people, making an Internet search, or checking through some books. Do whatever is required to assure yourself that a product incorporating your idea would truly be novel. It would be foolish to try to reinvent the wheel, and it would be just as foolish to self-evaluate the commercial merits of your new product concept.

Lawyers don't take themselves on as clients, and surgeons don't operate on loved ones. Both professions understand that where personal emotion is involved, good judgment often flies out the window. Why then do inventors think that they can objectively evaluate their product inventions? Does anything involve more personal emotion than the creation of an idea?

Why do so many inventors still spend money on books and software that claim to help them determine if their product idea will sell? No matter how many checklists they fill out, the result is always going to be a foregone conclusion ... the idea is brilliant! I'm not suggesting that the folks who write these books or develop this software are being deliberately deceitful—it's just that the facts speak for themselves. These books and programs don't work. They never did and they never will.

Almost as bad is the habit of relying on the praises of family, friends, and patent attorneys, all of whom have a vested interest in saying what you want to hear. Friends and relatives want to please you, and *no* patent attorney I've ever met will volunteer that your idea stinks. Why should he? What merchant tells a customer not to buy? And invention submission companies—forget it! Naturally they'll tell you your idea is brilliant. How else can they sell you all of their services?

I appear on lots of radio talk shows, and the host almost always asks what the inventor's first step should be when she has an idea. My advice to them is first, to make darn sure that you have a truly original idea, and second, to go immediately to a business expert—someone with no vested interest in you or the success of your idea—to see if it really has merit. And if he tells you it doesn't and he has the credentials to know what he's talking about, my best advice is to drop the idea and move on. Don't waste your time or your money. Attempting to prove the experts are wrong works once in a while—but it's usually just an expensive, wasteful pursuit. If you can come up with one idea, surely you can come up with another.

Also, in doing the research and consulting the expert, you must understand that "Will it sell?" is not the same question as "Can I get it licensed?" Trust me, they're not the same thing. In affluent America, where we're awash in discretionary income, almost any reasonably decent idea, if well designed and attractively packaged, will sell. In America, manufacturers blithely make and sell some of the worst junk that you can imagine. However, when it comes to handing out royalty checks, the standards are much, much more strict.

In order to interest a prospective licensee in a new product concept, not only does it have to be thoroughly designed and proven to work, but it must also be novel, it must be exclusive, it must represent an obvious advantage over existing products, it must hold the promise for a high volume of sales and profits—and finally, it must have the kind of impact that will cause a prospective licensee to say, "Wow! How come nobody thought of this before?"

The hard truth is that companies *hate* to sign licensing agreements. They don't like the constrictions that a licensing contract places on them, they don't like the bookkeeping expense of reporting specific sales, and they *really* don't like paying out royalties to outsiders. They'll go for a royalty deal only if the idea presented to them is really new, terrifically exciting, and holds out the hope of big sales and profits. Lots of mediocre products that were dreamed up internally and thus don't require royalty payments are brought into the marketplace every day— but often a different set of rules applies when it involves licensing agreements and handing out royalty checks.

That's not to say that inventors of simple, ordinary products aren't commonly striking deals that make them wealthy, but you must develop a proper perspective about the realities of the marketplace and the unique requirements for licensable new product ideas. And I can't stress the following point too much—don't be concerned if your idea seems too simple and obvious. Sometimes they're the best ones if they strike a responsive chord with the licensee. I'm almost embarrassed to tell you about some of the really foolish ideas I've sold.

My First New Product Idea

When my manufacturing business failed and I decided to earn my living by developing and licensing new products, my first target was a small giftware company located near my house. The reason for this decision was simply that I was driving a junker car and was afraid to travel too far with it, and paying for an airplane ticket to see a distant manufacturer was out of the question.

This company imported and distributed artificial flowers, so I obviously had to develop something in that category. What I finally came up with was a single silk rose attached to a greeting card and packed in an open gift box with a clear plastic lid. The cards, as I recall, said things like "A Rose for My Sweetheart."

I called the company, got through to the president, and made an appointment. I can still recall parking at the far end of the lot so that no one would see the car I was driving. As nervous as I was, I must have made a decent presentation because the president loved the idea. We made up a simple agreement on the spot, and I floated out of his office with a $4,000 advance check in my pocket. In short, I was paid thousands of dollars for taking an artificial rose, sticking it on a greeting card, and putting it into the bottom half of a hankie box. And the president actually thanked me for giving it to him!

It's hard to imagine a dumber, simpler idea, but the roses actually sold quite well, and I earned royalties from the product for several years. It wasn't exactly a million-dollar idea, but it was enough to convince me that licensing was a legal way to coin money. What I learned was that by presenting a product properly, it is possible to sell almost anything. The president didn't say the idea was too stupid to be believed. He recognized that the company could make a profit from it and was happy to pay me my small share. That kind of insight is what makes businesspeople successful. They understand the old saying that money never started the idea. It's the idea that starts the money.

The Zero-Sum Problem

Although that idea was not one that I'd boast about, it passed one crucial test that inventors often overlook and that can usually be determined by an understanding of the marketplace. What my flower idea did was create new sales rather than simply switch sales from an old product to the new one. Otherwise, why bother? If the end number of sales will be the same anyway, why would a manufacturer switch from a product that he has free and clear to one that he has to pay a royalty for? It may be a way for you to collect royalties, and it might even provide the consumer with a better product, but what's in it for the manufacturer? Remember, as I previously pointed out: Your customer is the manufacturer—let her worry about the consumer.

A few years ago a friend of mine designed a new picture hanger that was obviously an improvement over the ones that already exist. I knew what the outcome would be, but as a favor I presented his idea to almost every company I could find that included picture hangers in its product line. As I was sure would be the case, everyone turned it down. "Sure," they said, "it's an improved product, but so what? Why should I pay money for molds for a product that won't increase my profits?"

These gadget companies don't sell individual products to retailers; instead, they sell them their line (of which a picture hanger might be just one of dozens of products). The retailer buys from one company rather than another because he believes that company has better prices overall or better packaging or better terms. The design of the company's picture hanger is not part of the decision process, and no retailer would switch sources because of it. Furthermore, the retailer is going to sell the same number of picture hangers regardless of the design, and the consumer doesn't shop from store to store to look for the best picture hanger; she simply buys what's available in the store that she's in.

So, as you can see, as nice as my friend's idea was, there was no incentive for anyone to license it. Would it sell if it were nicely packaged and hanging from a peg in a store? Sure it would, and the consumer might even notice the improvement, but sadly I must say ... so what?

Before investing further in your idea, at the very least make sure it's significant enough to warrant a licensee's attention. I'm always sorry when I have to turn down product ideas because they're simply not important enough to bother with—new designs for toothpick holders or napkin holders, for instance—because I know that with a minimum of research the inventor would realize that the market is simply too small to turn the head of any prospective licensee.

As in Life, Timing Is Everything

If somebody says, "How come nobody thought of this before?" that should be music to your ears. The trick is *not* to develop something that's years ahead of its time. A big payoff results from creating something

Not Every Product Idea Has Commercial Merit

that's about *15 minutes* ahead of its time. And you'll really hit the jackpot if you come up with something that's about *5 minutes* ahead. Like a surfer looking for the perfect wave, if you time it just right, you may be in for an exciting ride.

As you know, for fad items, timing is everything. "I have the next Pet Rock!" people tell me, with visions of moneybags dancing in their heads. They're chasing yesterday's dream product in a world that has passed it by. Times change, people change, tastes change. A fad product that was perfect three years ago might be a flop if it were introduced today. The Pet Rock was a lucky fluke that comes along every once in a while. Maybe you'll be the lucky person to come up with the next one. It's not likely, but it's possible.

How a Filthy Mouth Opened Purse Strings

A few years ago, a friend of mine, Budd Goldman, created and marketed a product called The Final Word. It was about the size of an electric shaver, and pushing a button activated an electronic voice that pronounced vulgar epithets through a built-in speaker. The worst, most crude swearing you can think of came out of this little box when the user pushed the button. Class, huh?

The Final Word was about as ludicrous as any product I've ever seen, and I've seen a lot of absurd products. If I had been asked to invest money in it, I would have told Budd to take a hike. But he was too smart to ask for my money or my advice. The product was a blockbuster of a hit, perfect for its time and perfect for its place. A little sooner or a little later, and it might never have gotten off the ground, but that's the risk Budd was willing to take. He has a high-wire marketing company. He has more flops than hits, but the hits are usually big enough to more than make up for the failures. Nobody with a new product idea is ever turned away from his door without an audience.

"Widgets! Widgets! Get Me More Widgets!"

Most of the companies that might license your product idea already have creative staffs who presumably do nothing all day but dream up new products. You might think that they're looking for the next hula hoop, but actually they're probably creating new versions of the products the company already manufactures. These are called "line extensions," and they are just what the name implies. If the company produces a line of 30 types of widgets, the research department is working on ways to extend the line to 40 widgets—new widgets, improved widgets, but always widgets. As the headline of a recent computer company ad said:

AT MOST COMPUTER COMPANIES,
R&D MEANS REPLICATE AND DUPLICATE

So when you come along with something really new that can actually be sold right next to the widgets, the company thinks you're terrific. And you may scratch your head and wonder why this big company never thought of your obvious idea.

At the other extreme, small companies like my friend Budd's don't have research departments at all. And they're so busy with the day-to-day running of their business that they have almost no time to think about new products. So if you present yourself as a source for profitable new ideas, you'll be welcomed with open arms.

Surprisingly, nearly all of the new products, for companies both large and small, arise from customers' suggestions because most firms are simply not structured to encourage creative thinking. They respond rather than originate, and that's why people like me can be successful. And hopefully, perhaps also people like you. Despite the huge research and development departments of many Fortune 500 companies and despite the millions they spend in the search for new products, the majority of patents awarded every year go to small, independent inventors. These people— dreamers, innovators, and creators—are the superstars of the business world. Even when a large company does introduce an innovative product, it can almost always be traced back to one freethinker in the company who refused to be denied. Corporate committees are geared to funnel profits effectively, not to invent the products that earn those profits.

"Hey! They Stole My Idea!"

The insatiable demand for new products never wanes. Thousands of new products are introduced every year, while thousands of others die quiet deaths. The cycle is as dependable as the tides. The supply never comes close to satisfying the demand, and yet the profession of dreaming up new products must be one of the least crowded in the country—there are probably 10,000 lawyers for every new product developer.

I urge you, therefore, not to worry about your idea having to be new and unthought of. It isn't new, and it was thought of dozens, or hundreds, of times before. What matters is that you're the one who's doing something about it. When your version appears on the market, all those other people will go around telling anybody who will listen, "Hey, they stole my idea! I had it years ago!" Whatever your idea is, it's going to be new to some company, and if it's good, the management will happily pay you for it. And they'll continue to pay you for it, perhaps for years and years, as long as it continues to produce profits. It might not be a million dollars, but it might be a cushy second paycheck.

The End of the Free Ride

We're approaching the hard part of the C.R.A.S.H. program. You've worked diligently to develop your idea. It's your marvelous brainchild. Now you have to look at it coldly and dispassionately, as if you were a stranger. If it's not made of the right stuff or if you're hesitant to spend the time or money to see it through, this is the go/no-go juncture. You have to make two honest appraisals: Is there true commercial value to your product? Are you willing to do what it takes to get it licensed? If you're not enthusiastic about either, this is the moment to bring the effort to a halt. No penalty, no foul. Just go back to the drawing board, confident that your next idea will be twice as good as the one you have now.

The more intelligent research you do at this juncture, the easier it will be ultimately to license your new product idea. If your new concept was a toy, for instance, and if you could produce a few samples to let kids play with, certainly a potential licensee would be interested in the results. If you took the samples to a day-care center and the kids really had fun playing with them, you would have a powerful sales point. On the other hand, if the kids showed no interest or were quickly bored, the idea might need additional work or perhaps should even be abandoned. Regardless of the results of your research, you'll be better off for having the information. This is the time to get it, before you reach for your checkbook.

There are a few basic criteria in determining the prospects for your idea, but the most important thing I can tell you is to be honest with yourself and seek the advice of professionals. The majority of new patents are awarded to private individuals, and an estimated 95 percent of them are never commercialized. The obvious conclusion is that many millions of dollars are wasted every year in paying for patents on product ideas that simply have no commercial value. But who's going to tell you your idea has no merit? Your lawyer won't, your accountant won't, the model maker won't, and certainly your friends and relatives won't. Why should they? It's something that you have to find out for yourself and from experts in the field who have no vested interest in your success or failure, and you must not delude yourself when the unfavorable facts are staring you in the face. If you have the money to waste, a patent makes an attractive wall hanging, but it's an expensive piece of décor.

"Don't Confuse Me with Facts"

In the adult school course I occasionally teach on new product development, the semester always starts with the students absolutely refusing to reveal anything about their new product ideas. Each person thinks the rest of us have been put into the classroom by the devil to steal his or her idea. By midterm, however, they loosen up and there's no stopping them.

One older student showed me his drawings for a new concept in portable display fixtures that he intended to patent and offer for license to manufacturers of trade show displays. I happen to know something about that industry and quickly saw that what he had developed was inferior to displays already on the market. Existing displays are lighter, easier to assemble, and more flexible than what my student had come up with. It's not that his idea was bad; it's just that it was behind the times. He did have a unique locking system for his display, but it was not necessarily an improvement over existing generic devices.

Gently, I tried to point all of this out to my student, but he still bristled at my comments. So I backed off. It was none of my business anyway. Because he so frostily ignored my evaluation of his idea, here are two scenarios, one of which surely happened when the term was over.

Scenario 1: The Bad News.   In this version, my student takes his idea to a patent attorney. The lawyer either knows the idea has no commercial merit or he doesn't, but either way, he keeps his mouth shut. It's not his business to give commercial opinions. Patents are his business, and he informs the student that he can probably get a decent utility patent on the locking device.

The student tells him to go ahead, and 18 months pass while the student waits with eager anticipation for the patent's arrival. The attorney proves to be correct, and finally he has the patent in hand. The student now sends form letters proudly announcing his new product idea to about a dozen manufacturers and naturally gets turned down every time (most don't even answer). Determined, he sends out another batch, and so on, until his list is exhausted. The episode is finished, and he's out somewhere between $5,000 and $10,000. Meanwhile he's angry at the stupidity of the exhibit manufacturers who wouldn't know a good product if they fell over it.

Scenario 2: The Really Bad News.   Instead of going to an attorney, the student goes to an invention marketing company that, after reporting that the inventor's idea is "brilliant," persuades him to buy a "market research" report to explore its potential. Shortly thereafter, (surprise, surprise), he receives an enthusiastic report about the thousands of trade shows held every year, the zillions of exhibitors who need booth displays, and how clever he is to have created such an exciting new product for this enormous market. The student comes down from the clouds just long enough to buy everything the invention marketing company is selling except a new suit for him to wear when he accepts his award at the genius banquet. As with the attorney, the marketing company, even if it knows, will not tell the student his idea is worthless (except to the marketing company). So it costs him about $15,000 this time, and he still thinks the manufacturers are stupid. If the student listened to me, I would have shown him what the Federal Trade Commission says about companies like this:

SPOTTING SWEET-SOUNDING PROMISES
OF FRAUDULENT INVENTION PROMOTION FIRMS

Think you've got a great idea for a new product or service? You're not alone. Every year, tens of thousands of people try to develop their ideas and market them commercially.

Some people try to sell their idea or invention to a manufacturer that would market it and pay them royalties. But finding a company to do that can be overwhelming. As an alternative, other people use the services of an invention or patent promotion firm. Indeed, many inventors pay thousands of dollars to firms that promise to evaluate, develop, patent, and market inventions. Unfortunately, many of these firms do little or nothing for their fee.

The Federal Trade Commission has found that many invention promotion firms claim—falsely—that they can turn ideas into cash. But, the agency says, smart inventors can learn to spot the sweet-sounding promises of a fraudulent promotion firm. Here's how to follow up if you hear the following lines:

"We think your idea has great market potential." Few ideas—even good ones—become commercially successful. If a company fails to disclose that investing in your idea is a high-risk venture and that most ideas never make any money, beware.

"Our company has licensed a lot of invention ideas successfully." If a company tells you it has a good track record, ask for a list of its successful clients. Confirm that these clients have had commercial success. If the company refuses to give you a list of its successful clients, it probably means there aren't any.

"You need to hurry and patent your idea before someone else does." Be wary of high-pressure sales tactics. Simply patenting your idea does *not* mean you will ever make money from it.

"Congratulations! We've done a patent search on your idea, and we have some great news. There's nothing like it out there." Many invention promotion firms claim to perform patent searches on ideas.

Patent searches by fraudulent invention promotion firms usually are incomplete, conducted in the wrong category, or unaccompanied by a legal opinion on the results of the search from a patent attorney. Because unscrupulous firms promote virtually any idea or invention without regard to its patentability, they may market an idea for which someone already has a valid, unexpired patent. In that case, you may be the subject of a patent infringement lawsuit—even if the promotional efforts on your invention are successful.

"Our research department, engineers and patent attorneys have evaluated your idea. We definitely want to move forward." This is a standard sales pitch. Many questionable firms do not perform any evaluation at all. In fact, many don't have the professional staff they claim.

"Our company has evaluated your idea and now wants to prepare a more in-depth research report. It'll be several hundred dollars." If the company's initial evaluation is positive, ask why the company isn't willing to cover the cost of researching your idea further.

"Our company makes most of its money from the royalties it gets from licensing its clients' ideas. Of course, we need some money from you before we get started." If a firm tells you this, but asks you for a large upfront fee, ask why they're not willing to help you on a contingency basis. Unscrupulous firms make almost all their money from large upfront fees.

The pity is that with some honest research, the student could have easily determined that his idea was not commercially viable. Unfortunately, even if he did find out, chances are he would have simply ignored the findings. His mind was set, and facts were irrelevant. After all, my student is no different from thousands of other inventors. It's no accident that so few patented ideas ever become commercial products—they don't deserve to be.

I've even seen top-flight executives, who should know better, spend thousands of dollars to research their own idea's potential only to ignore the negative findings. When personal egos are involved, it's very hard to be objective, regardless of who you are. But just as you fell in love for the first time, you have to be prepared to fall out as well. This is the moment to ask difficult questions and to heed painful answers.

IS YOUR PRODUCT MARKETABLE? IS YOUR PRODUCT LICENSABLE?

When I wrote the original version of this book years ago, I used this section to offer some questions for you to ask about your new product idea to determine its value before pressing forward. Now, having since been in contact with literally thousands of inventors, I have a much better way to use this space. The questions I used to ask are as useless as the checklists and books that cover the same topic. No inventor will be dissuaded by my questions, just as they won't be from the lists provided by others. I could ask you, as lists I've seen do: *Does your product offer a unique solution to a real problem? Is your product easier to use than existing products? Is there a growing market for your product?* But if I did that, who among you would say no? I can easily provide a list with dozens of questions like this and you'll nod yes for each one.

Instead, I'd rather suggest that if you've already invented the product, before doing anything else, make sure someone hasn't already thought of the same thing. If it's potentially patentable, you can go to www.uspto.gov (the PTO web site) and do a free preliminary search. I'll discuss this in more detail in the next chapter. If it's probably not patentable, that doesn't necessarily reduce its value, but determining its originality might be a bit more difficult, and some legwork on your part should provide the answer. If it's already on the market or was in the past, folks in that industry should be able to tell you.

The next step is to go to an outside professional to get an unbiased reading on your idea's merits. Inventors by the thousands send me their ideas, and I'm always surprised at how many I'm compelled to turn down simply because I know it's not a new idea. If I know it, they should have been able to know it on their own. I hate to write rejection letters because I understand the hopes and passions tied up in a new product idea—but inventors come to me for straight answers and that's what I give them. If the product already exists or existed previously, we can't license it, period. That doesn't mean it can't be marketed—companies knock off existing products all the time; however, it does mean that no one will pay you royalties for the idea.

On the other hand, if you have *not* already created or invented a new product but have a head full of ideas, I urge you to first study the marketplace. I realize that what many inventors like to do is invent, and that's fine if you want to view it as a hobby. However, if you want to make money out of it, you have to understand that inventing is a business just like any other. No manufacturer would enter a marketplace that she's not familiar with or hasn't researched, so why would you? It's not enough to simply dream up a product; when you show it to a manufacturer, you must be able to discuss it in terms that he'll understand and appreciate. As I previously discussed, your customer isn't the consumer; your customer is the manufacturer. The more you understand about his marketing needs, the nature of his competition, and the way he distributes his products, the better able you'll be to tailor new product ideas to his needs. Your job is to worry about pleasing him and to let him worry about pleasing the consumer.

If you've analyzed your product idea as objectively as possible and if you still like what you see, the next question is whether you're willing to do what it takes to get it licensed. It's not a free ride, but if you're willing to pay the fare, I think you'll enjoy arriving at the destination. So, if your research and good sense tell you it's an original, do-able, profitable idea and if you're committed to seeing it through, then it's time to move forward with the program.

IF THE IDEA IS SO GOOD, SHOULD I SPEND MONEY TO PROTECT IT?

Inventors in the true sense of the word, who have invented or discovered something truly profound that is destined to have a long life span, must, of course, seek patent protection. A strong utility patent can provide some peace of mind and serve as an excellent asset when negotiating to license an idea. More on this later.

On the other hand, if what you've created can best be described as merely a clever, commercial idea or novelty product, then it may neither merit nor need that kind of protection. The awarding of a patent does not necessarily have a direct bearing on the value of a product idea. If it warrants a patent, good sense says you should get one. But if it doesn't warrant one, that doesn't mean it's any less worthwhile. 3M's Post-It notes aren't patented. Wouldn't you like to be receiving royalties on those little guys? Most of my own ideas are in the Post-It notes' category (although not in their league), and my visits to a patent attorney's office are few and far between.

You Don't Have to Patent Everything That Moves

Years ago, I created a funny little product called Sip 'N Lips. Did you ever buy wax lips when you were a child? They're usually sold around Halloween and come as big, red lips for girls and silly looking fangs for boys. My new product concept was to make lips like these out of durable plastic instead of wax and to have a funny loop-the-loop plastic straw attached. The idea was to make it fun for kids to drink their milk or other beverages. It was a quick product to dream up and an easy one to license. It's still being sold, and I still receive royalties. It's quite popular, and I imagine it will remain on the market for years to come.

My point is that applying for a patent never occurred to the manufacturer or to me, although we probably would have been successful. It was never even a topic of conversation. By silent agreement, we decided the product did not warrant the expense and effort. More important, we also silently agreed that it would not be worth the financial outlay to defend the patent against any potential competitors. In other words, I never offered to patent the product, and the licensee never pressed me to. These are business decisions based on experience and the understanding of what a patent can and cannot do for you. For some products, it's worth it. For other products, it isn't. This particular product has continued to prosper without legal protection, as the manufacturer and I were confident it would.

As a rule of thumb, if your new invention is designed to be in use for a number of years and if it requires a substantial investment on the part of the licensee, she probably won't proceed unless the product is well protected with a strong utility patent. On the other hand, if your idea is a simple product like, for instance, a toy or gift item or a little novelty product, getting a patent might be a waste of a great deal of money. Licensees don't expect it, and it's usually not the determining factor as to whether they'll license it or not. Most of my product ideas are in this category, and I've licensed more than a hundred of them over the years. I can't say that I've licensed every new product I dreamed up, but I've never been turned down because of patent considerations.

"Wanna Patent? Just Pay the Cashier"

Sadly, the first thing many an inventor does after creating a product idea is see a patent attorney. It's not the attorney's job to tell you if your idea has commercial merit or if getting a patent is even going to do you any good. You want a patent? The attorney will get you a patent. Pay the cashier on your way out. I urge you instead to pause, take a few steps back, and examine your idea in a critical light. A patent will cost at least five thousand dollars, and maybe much more. If it's necessary, do it. Just don't do it automatically.

I could get a patent of one kind or another on virtually every product I've ever created, but I seldom do so. My product ideas usually don't warrant it. Sometimes, the licensee will get a patent in my name, which I then sign over to the company. I'll review this later when we look at a typical licensing agreement. For now, however, I want to stress that not having a patent shouldn't deter you from presenting your idea to companies. It's extremely unlikely that they're going to steal it—not because you're dealing with such righteous people but because it simply doesn't make good business sense.

Why Companies Won't Steal Your Idea

1. *You'll Sue.* If they steal your idea, they know you'll probably sue and win. You'll have legitimate proof on your side regarding the idea's ownership. Even if you don't have a utility patent or a design patent, you may have a trademark, a copyright, a nondisclosure registration number, or detailed, witnessed notes. You'll have something. They'll have to manufacture proof, perhaps requiring the cooperation of some employees, and that is risky business.

2. *You'll Take It to a Competitor.* If they were going to steal your idea, they'd have to do it when you made your initial presentation by

telling you it was an idea they already had thought of. Logically, you would then take the idea to their competitor. So what will they have gained? If they worked with you, they'd have a head start on a great new product. By cheating, they create their own competition. What's the point? It's just not worth it.

3. *You Can Be Bought Cheaply.* The per-piece royalty they have to pay you is quite small, usually 5 percent of the wholesale price, and that cost is passed on to their customers. It's factored in as part of the manufacturing cost, so it's cheaper (and safer) to pay you than to steal from you and invite legal problems.

4. *You'll Cut Them Off.* The most important reason of all is that if they attempt to steal your idea, you'll never bring them another one. And your next idea may be sensational!

Regardless of what you may see in the movies, companies aren't built by stealing ideas from starving inventors working alone in dingy basement laboratories. They may try to negotiate to pay us less than we want, but they're satisfied to pay us something. That's just good business. It's difficult for me to get people to believe this, but you're never going to achieve

Companies Aren't Lurking to Steal Your Idea

success at licensing unless you do. You can't simply stand guard over your ideas, and you can't try to patent every little concept you think of—only your lawyer will thank you for that.

The Heimlich Maneuver

Back in the late 1930s and early 1940s, second-tier studios like Universal-International and Republic made B movie after B movie with a recurrent theme: The kindly, gray-haired inventor, toiling selflessly in his basement laboratory, finally discovers (1) an incredible new military weapon, (2) a formula to cure hemorrhoids, or (3) a secret device to turn camel dung into hard, brilliant diamonds. Getting wind of the discovery, Otto Heimlich, the evil president of Trans-Global Megalith, sends henchmen to break into the inventor's lab to steal (1) The Secret Death Ray Machine, (2) The Secret Formula, or (3) The Secret Plans! The thugs are caught in the act, a tussle ensues, and the aged brave inventor is left (1) dead, (2) blind, or (3) in a blue funk. Grief-stricken, swearing vengeance, and hurling herself at the impenetrable gray walls of the vile conglomerate is sweet, pretty Mary Wilson, who is either (1) the inventor's daughter, (2) his niece, or (3) his ward. The evil president and his thugs laugh fiendishly at the futility of Mary's rage. Then, from out of nowhere appears a brave soul who rushes to her side. It is (1) the handsome, fearless detective, (2) the handsome, fearless reporter; or (3) the handsome, fearless patent attorney.

I won't bore you with the rest of the tale except to say that it represents an extremely odd phenomenon. Some evil specter implanted this story into the brains of naive, young inventors-to-be, where it became almost impossible to dislodge. Now grown, millions of inventors will swear that Otto Heimlich still exists, maneuvering to steal their discoveries and inventions.

Never mind that poor Otto would now be well over 100, needing a walker to lurk about. In their minds, he's still out there, ready to steal their inventions, and for sure they'll die, go blind, or go into a blue funk. When someone innocently says, "Tell me about your invention," you can see the panic flash in the inventor's eyes.

I beg you, please don't panic and freeze in your tracks when a company refuses to sign your nondisclosure agreement. (More on that later.) Most of them won't. You have to assume that companies aren't run by neo-Heimlichs, waiting to steal your idea. It's far more profitable for them to do business with you as an ally than to have you feeding ideas to their competitor.

If you're not comfortable going out to sell your idea yourself, that's OK; we're not all cut out to be marketing people. Instead, get someone like me to do it. That's my business. But whether you do it or someone else does it for you is beside the point. The main thing is to get your idea in play. If your product idea's good, if it makes commercial sense, some company out there will be willing to pay you for letting them market it. But first they have to see it.

Having said that, it would be naive not to acknowledge that there are still some unscrupulous operators out there who will try to steal from you. Common sense dictates, therefore, that you should give your idea an appropriate amount of legal protection. The following section discusses the options that are available.

WHAT YOU SHOULD KNOW ABOUT PATENTS, COPYRIGHTS, AND TRADEMARKS

You may already have a general idea of the distinction between the basic forms of legal protection, but before we proceed, let's briefly review them.

1. *Utility Patent.* This is applicable for the invention or creation of a basic new process, machine, manufacture, or composition of matter. The idea must be primary, useful, original, and operational. Utility patents are designed to provide 17 years of exclusive use of an idea. What they actually do is give the owner the right to sue others who may decide to make unauthorized use of the patent.

2. *Plant Patent.* Here's how a government booklet describes this type of patent:

   > The law provides for the granting of a patent to anyone who has invented or discovered an asexually reproduced and distinct and new variety of plant, including cultured spores, mutants, hybrids, and newly found seedlings, other than a tuber-propagated plant or a plant found in an uncultured state.

   The Patent and Trademark Office (PTO), which is part of the Department of Commerce, will usually require that this type of application, unlike other patent applications, be accompanied by an exhibit. If a patent is awarded, it's good for 17 years.

3. *Design Patent.* This patent offers protection for a unique, ornamental design of the exterior of the product. It has nothing to do with how it works, only with how it looks. For instance, Auguste Bartholdi, the sculptor, had a design patent on the Statue of Liberty. It's not illogical to have a utility patent and a design patent on the same invention—one covers how it works and the other, how it looks.

4. *Provisional Patent Application.* Since this book was originally published, this new program has been added by the Patent and Trademark Office, particularly for the benefit of independent inventors. This is not a patent per se. What it does is save your place in the patent submission line for one year—the intent being to give you the time to determine if the idea has sufficient commercial merit to actually apply for a regular patent. The PTO does not offer an official opinion on your idea, but it does offer a valuable option. If, within a year after applying, someone else attempts to steal your idea or arrives at the same idea independently and makes a patent application or if you determine to proceed with a regular application yourself, yours would take precedence over the latecomer's. You have no obligation to proceed, just as the PTO has no obligation to approve your application if you do. However, at least for a year, the government offers a degree of protection. This costs much less than a regular utility patent submission, and I'll discuss later when and how it can be used. Another important consideration is that you are legally entitled to put "Patent Pending" on any prototypes, drawings, or presentation material that you might prepare.

5. *Trademarks.* The U.S. Department of Commerce has a useful booklet about trademarks available to the public. According to the booklet, "A trademark may be a word, symbol, design, or combination word and design, a slogan or even a distinctive sound which identifies the goods and services of one party from another." For instance, Coke and Pepsi are trademarks.

6. *Copyrights.* Patents and trademarks are issued by the PTO. Copyrights, on the other hand, come from the Library of Congress, because copyrights offer protection for artistic and literary works. Books will merit a copyright, as will works such as movies, cartoon characters, drawings, plays, songs, and so on. If you'd like to have a government publication on this subject, look in the Appendix for ordering information.

You'll Be Happier If You Leave It to Experts

As I've previously urged, before running off to apply to get your idea patented, you should first exhaust your personal means to determine if you truly have an original idea. And then, assuming you do, consult with an expert to determine if it has sufficient commercial value to go forward and if it's the type of idea that warrants the expense of patent protection.

If you have determined that your idea does need a utility patent, I don't advise that you attempt to secure one without a patent attorney. Although you can do it yourself, the patent you secure on your own may not be as ironclad as one secured through a trained professional. It's an expensive process, but a good lawyer will enable you to get all the protection your idea merits. A weak utility patent will be almost valueless. I am a big believer in doing everything you can on your own without lawyers, but I would never file for a utility patent by myself.

There are several software programs available, and at least one well-known book on the market, that advise folks how to apply for a patent themselves. The author of the book I have in mind, a patent attorney, has an excellent reputation, and I know that some find his book to be extremely useful; however, the more I read through it, the more convinced I was that the best move is to leave it in the hands of experts. It's too important not to.

Lots of patent mills out there (you see their commercials all the time on late-night cable) can get one sort of a patent or another for almost any product idea, but later the inventor usually makes the painful discovery that not all patents are alike. What you want is a *strong* patent, one that gives your idea the kind of protection to which it is entitled. I suggest you work with a credible attorney whom you trust to have a full understanding of your idea and the skill to get you the maximum protection for it. I don't have the years of training to do that for my own ideas, so I personally would rather use an expert.

In an average year, about two hundred thousand patent applications are received, and a little better than half are granted. Also, the average waiting time to receive the patent is about 20 months. These are not wonderful statistics, considering that even the simplest application will probably cost several thousand dollars, with no upper limit. Nevertheless, if what you've invented is valuable, it deserves the strongest patent you can get. Settling for less might one day break your heart.

Apropos of this discussion, *Inventors' Digest,* the magazine devoted to the needs of private inventors, published its own "First Ten Commandments of Inventing" on its web site, which are as follows:

FIRST TEN COMMANDMENTS OF INVENTION

Reprinted with permission from Inventors' Digest—*www.inventorsdigest.com.*

1. Stay away from invention marketing companies that advertise on radio and late night TV. They're out to fatten their wallets and empty yours!!! (Remember to ALWAYS ask for references!)

2. Keep good records about your idea ... someday they may be the backup you need to prove YOUR idea is YOURS !

3. Go to a Patent Depository Library and do your own patent search, or do the search online at www.uspto or www.ibm.com/patents. If you find that your invention is already patented, there's no need to go to a patent attorney. For a list of libraries go to www.uspto.gov.

4. Build a model. No need to get fancy at first ... cardboard, white glue, balsa wood, off-the-shelf parts. No matter how simple the idea, prove it works.

5. Have your invention evaluated by a non-biased professional. (Even if your Mom's in the business, go to someone else!)

6. Read all you can about new product development. Go to your local bookstore or library ... others have gone before you. Don't reinvent the wheel. Go to the UIA Center Bookstore online at www.uiausa.org.

7. Network with other inventors. Join a local inventor's organization. Go to www.uiausa.org for a complete list of groups.

8. If your patent search looked promising (see #3), make an appointment with a patent attorney, patent agent or patent search firm.

9. Do what you do well and hire pros to do the rest.

10. Don't fall in love with your invention, but if you're really sure you've got a winner (see #5), hang in there! Even "overnight" successes take a while!

Use It or Lose It

If you do apply for a patent, there's no need to wait the two years until you receive your patent before showing your idea to potential licensees. When your attorney makes the filing, you should put "Patent Pending" on all your material and proceed to offer your idea in a vigorous manner. If you're going to sit around cooling your heels until you actually have a patent, your terrific product idea might quietly become passé.

"Patent Pending" gives you no legal protection whatsoever, because you may never actually be awarded the patent. However, it does warn everyone that an application has been made, and when and if the patent is awarded, you'll have the legal authority to go after anyone who has copied it. Competing manufacturers usually pay attention to this. They don't want to invest thousands of dollars in tooling to knock off a product that may become patented even before their knockoff is on the market. But it doesn't always work that way.

Some small manufacturers automatically put "Patent Pending" on all their packages, even though no patents are ever applied for. The hope is that it will deter the competition, but the irony is that because the competition does the same thing, neither one pays any attention to the warning. Of course, it's against the law, but it's a tough law to police. Large companies would never put false "Patent Pending" warnings on their packages, and even fast-buck companies will pause if large tooling costs are involved.

Design patents aren't usually worth the expense unless the design itself is absolutely critical to the success of the product. Then, it's worth its weight in gold. Here's an example of what I mean:

Three companies are the major producers of those little cardboard air fresheners that hang from cars' rearview mirrors. Companies A and B own the licensing rights to every hot property on the market. I mean properties like Playboy and Spiderman and NASCAR and the NFL teams. All that company C owns is a design patent on an air freshener in the shape of a pine tree. The presidents of companies A and B each told me independently that C is the leader in the business.

I mention this because it's a unique situation. For company C, the design patent is extremely valuable, the basis for its business. For some reason, the pine tree shape resonates with motorists—and it's the only air freshener company allowed to use that shape. However, it's rare that the ornamental design of a product is all that important. It's not difficult to circumvent a design patent, and one design is usually as good as the next. You must be the judge as to how this applies to your own new idea.

Great! (Or Just Better Than Nothing?)

Before we leave the general area of patents, I'd like to say a bit about the new Provisional Patent Application program. Attorneys may disagree on the program's merits, but they do agree that if your product idea warrants patenting and if you can't afford to apply for a regular utility patent, the least you should do is apply for a provisional patent application. A utility patent application can easily cost $10,000 and up, whereas it's no problem to find an attorney to file a provisional patent application for you for under $1,000—or, for a few hundred dollars, you can buy a software program that will walk you through the application process on your own.

What I can add is my belief that the program is more valuable if you intend to market the product yourself as opposed to licensing it. The main reason is that, as mentioned previously, it gives you the legal authority to put "Patent Pending" on your product, which will deter knockoff guys for a year and maybe permanently if you do proceed to successfully get a regular utility patent. In licensing, it has less value because the prospective licensee has no guarantee that you'll actually apply for a regular patent, and if you do, he has no guarantee that you'll get it. However, the provisional at least has some psychological value, and I'm of the school that says either way, it's better than nothing.

If It's from Your Brain, Copyright It!

Most of my own new ideas qualify for copyright protection. This is something you can certainly do yourself. In fact, your original work is deemed by the government to automatically have copyright protection as soon as you create it. Nevertheless, I automatically include the copyright notification on all my original work, and you should certainly do the same. You've seen it many times: It's the C in a circle with the date and the author's name (for example, © 2002 Harvey Reese). Look at the copyright page to see how carefully the publishing company protects the copyright for this book.

Also, although it's no longer mandatory, it's also a good idea to register your work with the U.S. Copyright Office. The government application form is included in the Appendix. The registration fee is only a few dollars, and a copyright registration gives you the right to go to court to protect your property. If your work is registered before any infringement occurs, you can sue for recovery of your costs, legal fees, and damages without the necessity of proving them. These are called *statutory damages*. Under the right circumstances, a copyright is perfect protection for your product. In the game of Monopoly, for instance, you couldn't get patent protection for the action of the game. After all, many board games involve throwing dice and moving pawns around the board. What you can protect with copyrights and trademarks is the distinctive look of the game. That includes the name, the design of the board, the cards, the money, the pawns—probably everything except the dice.

If you have questions or would like some additional literature about copyrights, you can call the Library of Congress at (202) 707-3000.

Name It and Claim It

Even if your product idea doesn't warrant a patent, perhaps you can give it a great name or slogan or look. This is the kind of material you can protect with a trademark, and it can potentially add enormous value to your concept. You can't overestimate the value of a catchy name that quickly and clearly tells what your product does. If you didn't call a hula hoop a hula hoop, what else could you possibly call it?

Sophisticated marketing companies pay a great deal of attention to the names of their products because they appreciate their importance. In fact, companies exist for the sole purpose of creating these names and command high fees for doing it. If the name is right, many manufacturers feel it's money well spent.

One of the best-known of these naming companies is Name Lab, Inc., of San Francisco. It is responsible for names such as Acura, Compaq, Geo Lumina, and Zap mail. The people who do this kind of work refer to themselves as "constructual linguists," and they call their company a "name development laboratory." It's quite serious work as you can see from the following paragraph about security, taken from one of Name Lab's pamphlets:

> As we routinely deal with such sensitive information as new products and corporate mergers, only Name Lab employees are admitted to our offices. We employ locked files, coded record systems, and a single copy documents policy. Reference materials are returned and project discs erased at the completion of a project.

Because companies spend so much time on this, it may pay for you to do it as well. If your product idea is just mediocre, you probably won't sell it. But if that mediocre product has a great name, a great slogan, or a great look, chances are you *will* sell it. It's that simple. Without the name, the Teenage Mutant Ninja Turtles would probably just be a couple of reptiles looking for a pizza.

Under the right circumstances, copyrights and trademarks are just as important as a utility patent and are just as vigorously defended. Toy trade magazines are filled with manufacturer warning ads telling other manufacturers to *not even think* about stealing their copyright-protected toys.

Knockoffs are frequent in the toy industry, and the lawsuits fly hot and heavy. It is unlikely that a more litigious group of companies exists anywhere. One of the reasons for the many infringements may be that the selling period for fad merchandise is so brief that it can come and go even before a trial date is set. The one who does the knockoff is often willing to risk either that a trial will never take place because of an out-of-court settlement or that, if there is a trial, the only penalty will be to pay a royalty to the patent or copyright owner. In the meantime, the offending firm perhaps has made a big score off the other company's creativity and can well afford to pay the legal costs.

The stakes are high, and with millions of dollars on the line, the knockoff companies factor the anticipated legal costs into the product's selling price. The consumers lose; the lawyers win. Fortunately, those of us who create and license the products are above the fray. As fiercely as the toy companies may compete with one another, I have never experienced or heard of an occasion where ideas were stolen from the people who create them. The industry pays out many millions of dollars each year in royalties and licensing fees and is happy to do it. Why not? Every dollar paid out represents about 20 dollars deposited to their own account.

Foreign Patents

Obtaining a foreign patent involves a long, expensive, complicated process, so you shouldn't attempt it at all unless your product idea has global significance. The United States is such a huge market by itself, and the cost to obtain and defend foreign patents is so onerous, that you should probably skip it unless you have a major sponsor behind you.

Most of the world's industrialized countries are joined by treaty in something known as the Paris Convention. Essentially, this treaty gives you the right, after filing in a member country, such as the United States, to file in another member country within one year, using the date of the original filing. For instance, let's say you applied for a patent in the United States. In the meantime, six months later, a French manufacturer learns of your product and decides to apply for a French patent. Subsequently, within the year of your U.S. filing, you also file for a French patent. Because both countries are signatories of the Paris Convention treaty, your application would take precedence over that of the French manufacturer. Conversely, if you never file outside the United States, the French manufacturer can copy your product and sell it at will throughout the rest of the world.

Various other patent organizations overlap the Paris Convention and also impact on the methods and time periods you have for making your patent application. These include the European Patent Office (EPO), the Patent Cooperative Treaty (PCT), and the African Intellectual Property Organization. Although in general I have suggested that you forego this type of protection, there are certain exceptions: If the U.S. company to which you license your product has a strong overseas marketing component, it might be interested in securing foreign patent protection at its own expense. The patent would be registered in your name and licensed to the company. Or if that mythical French manufacturer comes to you for a license, the licensing agreement could specify that the company will secure a French (or European) patent in your name, to be licensed to them. Finally, of course, if you have a foreign company ready to go, you might explore getting the patent on your own.

The Perils of Joint Ownership

If the product was invented jointly by you and your brother-in-law Melvin, you will file for the patent as co-inventors. The PTO doesn't care whether you did 90 percent of the work and Melvin did 10 percent. As far as the government is concerned, you are each equal inventors and can act independently of each other in selling or licensing your patent. It doesn't take much imagination to envision the problems that can develop from such an arrangement, as each of you could actually wind up competing with the other to license the invention and keep all the royalties. It's all perfectly legal. One alternative is to have a presubmission contract that spells out how the two of you will proceed. The second, and better, solution is not to have a partner at all.

It's a natural tendency for people to look for partners whenever they embark on a new venture. There's comfort in having someone to share the problems with, and it gives many people the courage to press forward. However, you have to confront the reality that almost every partnership eventually fails, usually with bitterness on both sides. Look at your own circle of friends, and you'll see what I mean. I've had partners twice in my career, and both times it ended unpleasantly, not because my partners were bad or dishonest people but simply because it's virtually inevitable that friction will develop when every decision must be made in concert. So if you really don't need Melvin's help, you're much better off by yourself. If you need companionship,get a pet.

HOW TO GET SOLID PROTECTION FOR YOUR IDEAS WITHOUT PAYING LEGAL FEES

First of all, keep your notes. They should be dated and filed in order. Then make a detailed drawing of your final design, date it, and have it wit-

A Pet Is Better than a Partner

nessed and notarized by two people who have no stake in it. Take one copy and safely file it away. Take another copy and mail it to yourself. Don't open the letter! For the price of a stamp, you will have a postmark that provides some evidence as to when you conceived the idea. Every patent lawyer I spoke to advised me that this old trick of sending a letter to yourself doesn't mean much in court because letters can so easily be tampered with. I understand this, but it's such an easy thing to do, and coupled with your other proofs, it adds to the preponderance of the evidence identifying a defining date of your invention.

Whatever Else You Do, Do This

Also, for a few dollars, you can file a disclosure form with the PTO in Washington, D.C. Just send your check along with a detailed description, drawings, or photos of your product design. Your pages should be numbered and no larger than 8½ × 13 inches. If you include a self-addressed, stamped envelope, the PTO will send you back a Disclosure Document Number. The department will keep your disclosure on file for two years. It is not passing judgment on the merit of your submission but is providing proof as to the date of its conception. A usable form is reproduced in the Appendix. The Disclosure Document Number, the witnessed copies of your drawings, and the sealed, post-marked envelope are like triple locks for your product idea, and the total cost will probably be less than ten dollars. It's doubtful that you'll ever need either one, but the cost and effort to obtain them is so modest that I feel you might as well do it—if only for your peace of mind.

You should also keep an accurate record of your expenses (including this book). If you act like an inventor and keep notes like an inventor, the government may consider you to be in the inventing business. As such, your expenses will probably be deductible by filing a Schedule C with your tax return. Your own accountant can guide you. Also, the Appendix provides a list of inventor organizations. You may wish to join one and avail yourself of the tax guidance and information that members receive.

The Paper Trail

Courts usually find that if you present a new and novel idea to a company—something that wouldn't be obvious in its normal course of business—the firm has an obligation to keep the information confidential and to not use it, or profit from it, without your permission. This is not hard-and-fast law, but judges will almost always find in your favor if you run into an unscrupulous company. To a large extent, it will depend on the nature of the information you disclose.

Suppose you saw a survey that said most women prefer turquoise over every other color. And suppose you went to a steam iron manufacturer with the idea that they change the color of the handles on their irons to turquoise. If a year or so later the company actually did switch to turquoise, that doesn't mean you could collect royalties on the sale of every steam iron. Your idea would never fit the court's definition of being new and novel. The color preferences of women is certainly something a company could be expected to determine in its normal course of business. However, if what you presented was an altogether new type of handle that could be switched around to accommodate either left- or right-handed users and if the company simply took your idea, the courts would almost certainly find in your favor.

To protect against such thefts, you must establish proof that it was your idea to begin with and that you introduced the idea to the manufacturer on a specified date. You do this by developing what lawyers call a *paper trail*. For instance, if you had a meeting with Franklin Smedley to show your new Switcheroo Handle, you'd send off the following letter after your appointment.

> Dear Mr. Smedley:
>
> It was very nice meeting you today. I appreciate your courtesy and am quite pleased with your initial favorable reaction to my SWITCHEROO HANDLE concept for your steam iron line.
>
> I am herewith enclosing the samples and drawings listed below for your internal discussion, and as you suggested, I'll call you in two weeks.

If Mr. Smedley said he did not like your new product idea, you should send off a letter like this:

> Dear Mr. Smedley:
>
> It was nice meeting you today, and I appreciated the opportunity to present my SWITCHEROO HANDLE concept for your consideration. I was sorry to learn that it doesn't fit in with your current marketing plans, but I certainly understand your reasoning.
>
> I thank you again for your courtesy and hope my next new product concept will be more suitable to your needs.

Mr. Smedley will understand that you didn't send him one of these letters just because your mother taught you such good manners. He'll realize that he's dealing with someone who knows the legal implications of the letter, and he will understand that it is like a warning shot across the bow.

And if Smedley calls you for some additional information, his mail tomorrow should include another letter from you:

> Dear Franklin (by now you're on a first-name basis):
>
> Thank you for your telephone call today. Now that I've had the time to reflect on your question, I'm more convinced than ever that the answer I gave you is correct. The framus will fit snugly on the thigmetz if you loosen the witzberg.
>
> Please call me again if I can be helpful. In the meantime, I'm still scheduled to call you next Thursday to see if you plan to proceed with my SWITCHEROO HANDLE concept for your steam iron line.

In other words, every contact you have with Mr. Smedley should be followed by a confirming letter, fax, or e-mail. Make sure you keep a copy for yourself. Also, keep a copy of your phone bills that reflect any long-distance calls made to Mr. Smedley or travel expenses to his office. If the need ever arises, you'll be able to completely document your dealings with that potential archfiend, Franklin Smedley.

For a modest investment, you can have letterheads and business cards printed or even do them on your own computer. This not only will make your dealings with Smedley more professional but may also prove quite beneficial when it comes time to pay your taxes. It would also be a good idea to set up a special checking account to pay any expenses you might incur—items such as travel, prototypes, artwork, and photography. I automatically create a paper trail, even with companies I've been dealing with for years. I've never had a problem, and I'm sure that it's partly because everything I do is documented. Gypsy Rose Lee said, "God is love, but have it in writing." If you hope for the best while you prepare for the worst, you'll never be caught off guard.

WHEN YOU REALLY DO NEED A LAWYER, HERE'S HOW TO FIND THE RIGHT ONE

If you decide you do need a lawyer, keep in mind that it's just another service business and you're a customer. Act like one. Shop, negotiate, ask questions. First of all, as obvious as this sounds, only use a patent attorney who knows what you're talking about. Patent attorneys, like doctors, often tend to specialize. If your invention involves complicated chemical matters, you probably will want a lawyer who's also a chemical engineer. Patent attorneys must be approved by the PTO in order to deal with them, so don't go to Uncle Bernie, who is a personal injury lawyer, just because he says he'll cut you a deal on the price.

Develop your list of potential patent lawyers from recommendations from friends, your regular attorney, inventor organizations, Internet queries, and the telephone book. If you still haven't found someone you feel comfortable with, the PTO can supply you with a directory of more than thirteen thousand individuals whom they have authorized to act on behalf of inventors. To qualify, the attorneys must have the appropriate legal training and also a college degree in engineering or a physical science. You can also use the services of a patent agent. His fees will probably be smaller, and he's fully qualified to prepare and file patent applications. What he can't do, because he's not an attorney, is initiate legal actions on your behalf or defend you if you have inadvertently trespassed on someone else's patent.

Don't Worry, Lawyers Don't Gossip

When you call an attorney to make an initial appointment, the very first question to ask is whether there will be a charge for the preliminary meeting. If you spend time having the attorney review your product, it's fair that she should charge you for her time, but if your conversation is limited to a discussion of fees, rates, and billing methods, no money should change hands. You shouldn't have to pay a fee for asking what the fee is.

How Long Is a Five-Minute Call?

In addition to learning the attorney's hourly rate, you'll also want to find out the minimum billable unit. Let's say you interview two attorneys, each of whom bills at $300 per hour. Attorney A has a minimum billable unit of 15 minutes, and attorney B's is 6 minutes. If you make a 5-minute telephone call to each of them, attorney A will bill you for $75 (¼ of $300), and attorney B will bill you for $30 (¹⁄₁₀ of $300). In my experience, most attorneys use the 6-minute system, so be properly alerted if you meet one who uses 15-minute units.

Because an attorney is going to bill you for *everything*, you must keep a log and organize your thoughts before making phone calls or keeping appointments. Plan your meetings. Have a list of the questions you want answered and be concise and to the point. It's been my experience that attorneys make great listeners and will gleefully laugh at every joke you tell, but it's on your time. Talk may be cheap but not when you're dealing with an attorney. The clock starts ticking as soon as you walk through the door and doesn't stop until you exit. So do it as quickly as you reasonably can. And be sure you clearly understand what a procedure is going to cost before your attorney undertakes it. There shouldn't be any surprises. This is simply common sense and should not cause a professional to take offense.

Beware of the "G" Word

If you call in a building contractor to remodel your basement, you'll receive an estimate for material, labor, and fixtures as well as a tentative schedule. You'll compare several contractors and finally make your choice. Dealing with a patent attorney is quite similar, and you should use similar judgment. Ask the several attorneys with whom you meet to break down their charges, which they'll be happy to do, and then you can make an intelligent decision. These are the likely components:

1. The search.

2. Government filing fees.

3. Drawings (number of views required).

4. Out-of-pocket expenses.

5. Professional services.

The fees quoted for the first four components will not vary much from attorney to attorney. They shouldn't because they all represent services the attorney purchases on your behalf. However, on the fifth component (professional services), you'll receive an incredible range of charges. For instance, if you check with four attorneys, the highest quoted cost for professional services may actually be twice as high as the lowest quoted cost. When I ask why this is so, nobody seems to know. No one would dare suggest that the higher price will get you a correspondingly higher level of service. After all, it's only a simple patent application. Could it be that they're afraid to say the "G" word? Could be that greed has something to do with it?

One attorney told me that she used to work for a prestigious patent firm, which prided itself on being the highest-priced firm in the East. If another firm raised its hourly rate to match those prices, this firm would automatically raise its rates even higher.

Although I'm not suggesting that you automatically give your work to the cheapest attorney, I am submitting that higher cost does not necessarily relate to a correspondingly higher quality of work. When all is said and done, it's still a judgment call, and if the fees are within your budget, you should deal with the attorney with whom you're most comfortable. Both you and the attorney hope that this is the beginning of a long relationship, so it's important above all else that you like the individual's personality and respect his ability.

The first patent attorney I ever used was hired by price alone, which turned out to be a big mistake. His hourly rate was about $20 less than the nearest next quote, but I probably overpaid by $1,000 before I was finished because he was inept and greedy. I say he was inept because he had to "research" simple facts, thereby wasting an hour here and an hour there. I say he was greedy because he padded the bill with petty charges. His secretary called once to ask for my birth date to put on a form. The call lasted about 25 seconds. My monthly bill reflected this as a half-hour "telephone conference with client." Obviously, I never used him again, and I learned a few years later that he was no longer with the firm. Fired, I hope. Because you couldn't humanly make all these possible mistakes on your own, you might as well learn from some of mine.

It Never Hurts to Ask

While we're on the subject of fees, if the price quoted is more than you can afford, it's quite appropriate to ask if the attorney can come down a bit. Please don't be embarrassed. It's a legitimate question, and you may be pleasantly surprised at the results. Lawyers are, after all, in business, and they don't want to see you walk out the door any more than a retailer wants to see you leave the shop empty-handed. Once the fees are settled, what about the terms? If you can live with the attorney's regular payment system, fine. If you can't, that's also negotiable, and I promise you won't be the first client to get better prices or extended terms.

Money considerations aside for a moment, should you use a large firm or a single practitioner? The large firm has specialists in virtually every discipline, so no matter how complicated your concept is, someone there will understand it. If the single attorney gets stuck, to whom does he turn? On the other hand, you're probably such a small fry for the large firm that you'll wind up with a youngster who still hasn't had time to get her law degree framed. The single practitioner may not have a backup staff, but perhaps he has years of experience under his belt. On the other hand, is his practice so small because he's not very good?

Would You Like to Be a Loss Leader?

My own instinct usually leads me to a medium-size firm. Patent firms tend to be much smaller than the giant corporate forms, so "medium-size" in my mind means approximately a six-to-ten-lawyer group. On the rare occasions when I use a patent attorney, I stay away from the extremes. I feel the single practitioner may be trying to do too many things alone to take care of my needs properly, and I don't want to be the big company's smallest client. A senior partner with a large firm recently confided that they lose money with individual inventors. They take them on only because they hope they'll grow into larger accounts. That probably represents the thinking of most large firms, which reconfirms my feelings that I'm better off with a smaller group. I don't know about you, but I don't want to be anybody's loss leader.

Would You Like Your Lawyer as Your Partner?

Sometimes, if your concept is patentable and particularly appealing, an attorney may be willing to provide services free in return for a percentage of your potential profits. Because of having a proprietary interest in your success, the attorney might be willing to provide leads and introductions and could give advice and lend expertise to the negotiation and contractual dealings with the licensee.

Some attorneys tell me that such an arrangement would violate the code of ethics of the American Bar Association (ABA). The reasoning is that an attorney presumably has much more knowledge of patents and licensing than a layperson; hence, an unscrupulous professional can take unfair advantage of a naive inventor. There's a great deal of merit to this reasoning, and you should be fairly warned. I know for a fact, however, that some professional inventors do have such an ongoing relationship with their attorneys. If the attorney is trustworthy, as most are, and if the inventor is sophisticated, this kind of relationship could be mutually beneficial. If the stakes are large and require substantial legal work, this option is worth considering. The decision depends on your own financial situation, the complexity of the idea, and the degree of confidence you have in yourself. Some attorneys will be willing to discuss an arrangement like this, and others won't. If you're interested, it can't hurt to ask.

What Happens When You Meet with Your Attorney?

Before your first visit, you must take the time to clearly write out the nature of your invention, supporting the description, if possible, with sketches and a model. You want to be sure that your attorney clearly understands from the outset the exact nature of your invention, its basis, and the existence of any past or present competition. You want to be thorough, and you want to be clear, but you don't want to make an oration. Remember that the clock is ticking, so have everything in order before you enter the office and stick to the point while you're there. In conducting research for your invention, you will have undoubtedly discovered information about other products similar to yours that may be on the market now or were in the past. Whatever you know that's pertinent should be conveyed to your attorney to save time and aid in research on your behalf.

Your attorney, after carefully reviewing all your material and possibly conducting some preliminary research, will tell you one of three things:

1. *Your Idea Is, er, Shall We Say, Prehistoric.* In the attorney's judgment, your idea isn't workable or is not sufficiently novel to earn a patent. As with a doctor, you can go elsewhere for a second opinion, but I don't recommend it. If you are told that your idea isn't patentable, I suggest you should accept that as fact. There's no incentive for an attorney to tell you that if it isn't true.

2. *Your Idea Is, Dare I Suggest, Pedestrian.* The attorney may advise that although it is possible to obtain a patent, it won't have much value because your invention isn't very profound. Lawyers boast that they can get a patent of one sort or another for almost every invention placed on their desk. They call these "red button" patents because, if nothing else, they can get a patent for the red button you

Your Idea Is, er, Shall We Say Prehistoric?

used to start your gizmo. If all they can get for you is a weak patent, you might as well save your money. Most lawyers will volunteer this information. If they don't, it's your job to ask.

3. *Your Invention Is, How Shall I Put It, Fantastic!* The attorney, after saying that your idea appears to have considerable merit and that getting a broad, solid patent for it should be possible, will undoubtedly recommend a patent search to see if someone else has beaten you to the punch.

Whatever opinion the attorney gives you about your idea, bear in mind that the evaluation is strictly a legal one and that the attorney is not necessarily passing judgment on its commercial value. Just because the attorney may say the idea isn't patentable does not have to mean that the idea won't have considerable commercial value to the right company. Conversely, if the attorney says she can get an iron-clad patent for your idea, that still doesn't mean the marketplace is anxiously awaiting its introduction. In other words, the attorney's expert legal opinion is neither cause to despair nor rejoice; it's only a step along your way.

THE PATENT SEARCH

There's no law that says you have to conduct a search before applying for a patent, but it's so sensible to proceed in this manner that everyone does it. A search is much less expensive than a patent filing, so it's logical to first make sure that your idea hasn't already been taken. And if it has, when you see the existing patent, you may be able to redesign a way around it.

Even before engaging a patent attorney, you should at least conduct a preliminary patent search on your own either by visiting the PTO web site or actually going to one of the patent repositories located in most major cities. The benefit of the latter is that the libraries have trained staff to help you with the search. If you find that existing patents preclude you from going further, you'll have saved the legal fees. However, if you don't find anything, that doesn't necessarily mean other patents don't exist. It might mean that you didn't know where to look. That's why you need a professional searcher. You can hire one directly or through your lawyer. It'll cost more through the lawyer, because he'll add a markup, but that might still be the better choice because it'll give him full control and full responsibility.

Hiring a Professional Patent Search Firm

If you don't trust yourself to conduct a thorough patent search—or if you want to be doubly sure before proceeding, it's not a bad idea to retain a professional searcher. Spending a few hundred dollars at this juncture may save you many thousands with an attorney.

You can find patent search companies in your telephone directory, through an Internet search, by networking with other inventors, and from inventor organizations. A visit to your local telephone company office will get you access to a Washington, D.C., classified directory where you'll find all the names you'll need. To give you an idea of costs, a professional searcher charges in the range of $75 per hour, and it will take the searcher five hours on average to complete a search on a simple mechanical patent.

One note of caution: In using the phone book, it's easy to confuse a patent search firm with a patent marketing firm because they both offer searches. Make sure you know which is which when you make your choice.

Applying for the Utility Patent

Assuming that the search uncovers nothing to conflict with your concept, your lawyer will now open your file and apply for the patent. You can now properly identify all your prototypes and presentation material as patent pending, and you should be proceeding with your licensing efforts while your application is going through the approval process. Be assured that the application for your patent, even if it's rejected, is held in strict confidence by the PTO. The public can never see it. Once a patent has been issued, however, anyone can get a copy by sending in the patent number and a nominal fee.

Without taking you through the application step by step, in essence, you will give your lawyer the power of attorney to act on your behalf in filing a standard patent application with the PTO. The attorney will clearly and precisely explain your invention, supported by drawings that have been rendered by a patent draftsperson who is familiar with the style and manner required by the PTO. Your attorney will explain what your idea is, what it does, and how it advances the state of the art in the invention's field. Up to this point, the work could probably be done by a paralegal. Where your attorney earns a fee is in the statement of the claims.

The First Action

The claims, listed one after the other, are what are proposed to be unique and exclusive to your invention. These are the points that you're asking the PTO to give to you and you alone for the next 17 years. Your attorney is going to attempt to make these claims as broad as possible, whereas the PTO examiner will attempt to keep them as narrow as possible; this is the difference between a good and a bad patent. If the patent is too narrow, competitors can more easily bypass it. If it's too broad and sweeping in its exclusivity, your patent is more valuable and will, perhaps, greatly improve your ability to license your invention.

Your claims will be reviewed by an examiner who will search through the appropriate existing and expired patents and available literature to ensure that your invention is truly new. The examiner's decision, called a *first action*, will be sent to your attorney:

1. The examiner may conclude that your idea is not new or novel and may therefore deny the patent.

2. The examiner may accept all the claims as submitted, and your application will be passed through for patent issuance (assuming you pay the appropriate fees).

3. The examiner may accept some of the claims you make for your invention and reject others.

If you conducted a search beforehand and if your idea passed through your attorney's informal screening, there's a good chance that it won't be rejected out of hand. There's also a good chance that your claims won't be accepted completely as presented. The most common action is to allow some claims and not allow others. Your attorney, after determining the case that can be made, will respond with evidence to support the original claims. The examiner may revise the decision or not, and the correspondence my go back and forth several times. At some point, however, the examiner proclaims the decision to be final, and the last course open is an appeal to the patents commissioner. The examiners' decisions are rarely reversed, and this final step should only be considered if the stakes are high and if you feel you have an exceptionally strong case to be made. Keep in mind that it's a time-consuming process and you're paying your attorney by the hour.

Assuming all the claims are eventually settled and the patent is issued, you will be expected to pay an issuance fee plus maintenance fees at various intervals. A fee schedule is included in the Appendix. You won't receive an invoice from the government, so it's up to you to know that these fees are due within $3\frac{1}{2}$ years, $7\frac{1}{2}$ years, and 11 years from the date of the patent's issuance. If you don't pay the fees, your patent will expire prematurely.

However, before you even retain an attorney, it's worth the few dollars to get the government booklets on patents, trademarks, and copyrights. I've merely touched on the subject here, and the booklets are quite comprehensive. They can help you determine how much you can do yourself. It's probably a great deal more than you think. And even if you do engage an attorney, the more you know about the process, the better a client you'll be.

Invention Marketing Companies Will Break Your Heart

As I mentioned, if you look up *Patent Attorneys* and *Patent Searches* in your telephone directory, you'll probably also notice listings for the dreaded invention, marketing companies (sometimes called *invention submission companies*). Most of them will take your money and run like a thief in the night. Such companies prey on novice inventors by offering market research, patent search, patent applications, manufacturer submissions, and license negotiations. I've had no firsthand experience with invention market companies, but I've heard and read enough horror stories to strongly advise you to stay clear of them.

First, you submit your idea for a "free" evaluation. You'll naturally be told that your idea is brilliant, and for the "modest sum" of seven or eight hundred dollars, they'll agree to do market research on the sales potential of your idea. Naturally, the report raves about the supposedly enormous market for your invention and implies that only a genius could have created it. That the report is canned, gleaned from thousands of other previous reports, seems almost beside the point. The victim is now excited and hopeful enough to willingly pay thousands of dollars for all the other services. Finally, when the string runs out and there are no more services to be sold, the inventor is a little wiser and a whole lot poorer for the experience. It's not uncommon for an inventor to spend $15,000 before she realizes that the results are worthless.

A California invention company, when forced to open its books by the state's attorney general, revealed that of all the thousands and thousands of people who paid for services over the years, only three individuals made a profit from the association. In the spirit of fair play, I offered to interview a principal from one of these companies to hear his side of the story—the only answer I received was a runaround. The man I spoke to (most of these are franchised operations) said he was "too busy" to see me. I asked if he could at least send me some of his literature. He promised he would, but, of course, nothing ever arrived.

I don't mean to suggest that these companies are necessarily operating illegally, because usually they're not. In a literal sense they do exactly as they promise. The problem is that it's all smoke and mirrors, and as the money is gently extracted, the inventor's hopes and dreams are seldom, if ever, realized. If you can't resist the siren call, at least ask the company to give you the names of at least six current clients. They won't do it, but ask.

If all else fails, take a few moments to compare what the company is offering with the following red flag warnings published by the United Inventors Association (UIA) on its web site and reprinted here with its permission:

UIA RED FLAG WARNINGS

Copyright © 1992–2002 Bob Lougher, Executive Director, UIA

The following list can be used as a guide to verify the credibility of an invention promotion company.

- Company refuses to provide in writing the number of ideas they have represented and how many inventors made more money than they invested.

- Company refuses to provide in writing the number of ideas that have been sent to them and how many they accepted.

- Company refuses to provide inventor with at least three clients (preferably in the inventor's own local area) that can verify their credibility.

- Salesmen apply pressure to send money in right away.

- Company tells you to fully describe your idea in writing and then tells you to mail this information to yourself and not open the envelope. This ploy is used to give the inventor the false impression that the idea is somehow protected. In fact, it does absolutely nothing.

- Company recommends that a design patent be applied for.

- Company provides a patent search without a patentability opinion.

- You can never directly reach the salesman without leaving a message. The salesman is most like working out of his home and is using a phone drop.

- Company claims to be located in one State but all correspondence is postmarked from another State. These companies commonly use fictitious addresses and mail drops to hide their true location.

- Company runs slick ads on radio, television, and in national magazines.

- Company offers a money back guarantee if patent does not issue.

- Company recommends submitting your idea to manufacturers without applying for a patent first. Most manufacturers we have talked to will not consider an idea from the outside unless it is at least patent pending.

Invention marketing companies should not be confused with backers, venture capitalists, and licensing agents. These people will sponsor or represent you in one way or another because they believe in your idea. Their remuneration is a percentage of whatever royalties your idea generates, and if it's not successfully licensed or marketed, they get nothing. Invention marketing companies, on the other hand, earn their money from the fees you pay them for "services" rendered.

Don't Forget about Angels

If you have confidence in your idea but lack the funds to retain an attorney to see it through, you might want to give some thought to getting a backer. Not a partner, a backer. In exchange for funding the legal and miscellaneous costs, the backer receives a portion of the net profits. The angel is not a partner—it's your invention, and you call the shots. The percentage for the backer is negotiable. If I were doing it, I'd offer my backer 25 percent, and I'd be willing to go up to a third. That's purely subjective, and your input is as valid as mine. The backer would be investing in you in the same way an angel invests in a Broadway show. Neither has a voice in the management, and the only time they should make an appearance is when there are profits to divvy up. If my circumstances dictated the need for financing, I would have no qualms about trying to arrange this kind of deal. Keeping 100 percent of nothing is still nothing.

In making our way through the C.R.A.S.H. course, so far we've looked at ways to help you create your million-dollar idea, we've explored ways to determine its value, and we've examined legal and practical

Angels Come in All Shapes and Sizes

ways to protect it. So far, it's been an uphill fight. Now begins the payoff process: Your new idea is ready to meet the world. In the next chapter, we'll discuss how to prepare it for its debut.

It's time for the "A" in our program. It's time for Action. Now we go on the attack.

4

SWING INTO ACTION

Prospecting, Getting the Appointment, and Preparing the Presentation

*Before anything else, getting ready
is the secret of success.*
—Henry Ford

About sixty-five years ago, so the legend goes, a teenage beauty named Lana Turner was sipping a soda in a drugstore at the corner of Hollywood and Vine when a big-time director came up and offered to put her in pictures and make her a star. That probably was the last time in history anyone ever succeeded by simply being discovered. Today, you have to push and hustle and yell, "Look at me!" if you want to get anywhere. A mediocre idea that has been successfully licensed is a thousand times better than a brilliant invention that's gathering dust on a shelf waiting to be discovered. As an old saying reminds us, Nothing happens until somebody sells something.

In referring to the idea for his stores, Frank W. Woolworth once remarked that he was the world's worst salesman and, therefore, he had to make his stores a place where it was easy for people to buy. You may feel you're ready to challenge Mr. Woolworth for that worst-salesman title, which is exactly why you must spend the time to prepare a professional, persuasive presentation. Think of it as your store, and like Woolworth's, its purpose is to make it easy for people to buy. No matter how sophisticated your potential licensee might be, he will always pause longer over a beautifully prepared presentation than he will if it's an amateur one.

HOW TO SHAPE UP YOUR IDEA FOR PRESENTATION: IT'S NOT IMPORTANT UNLESS IT LOOKS IMPORTANT

As the scene opens, showing the interior of a chic Park Avenue café, you are seated at a quiet table with Max Dubois, president of Global Amalgamated Things, Ltd. "Max," you say, as the waiter deposits the second round of perfect Roy Roys, "I have this little idea to show you. I think you'll find it amusing."

You take your Meisterstuck Mont Blanc pen from the inside pocket of your Yves St. Laurent jacket and quickly sketch a simple design on a cocktail napkin. The ink smears on the damp paper, but the design is still barely legible in the dim café light. With a look of greedy anticipation on his face, Max deposits his Havana Luxo in the ashtray, puts down his drink, leans over, and, with pudgy, trembling fingers, extracts the napkin from your hand.

The second hand takes several laps around your Rolex Oyster as Max studies your drawing. People pass by the table and extend greetings to him. He simply waves them off, his beady eyes never leaving the napkin. Finally, he mutters half aloud, half to himself, "Genius. Pure genius." He reaches into his pocket, extracting a Hermes eelskin wallet and checkbook holder. "I must have it," he says. "Just name your price. I must have it."

Fade to black.

In real life, you'll probably never go on the town with Max, let alone share perfect Rob Roys with him. But if you're going to sell a new product concept to his company, chances are good that he'll see your presentation. And he's not going to buy if it's on a cocktail napkin. So creating a professional presentation for your idea is as important as any other single step in the selling process. The key to success is self-confidence, and the key to confidence is preparation.

Dreams Don't Open Checkbooks

The problem is that you're selling a dream and your customer doesn't buy dreams but rather invests in reality. That's how the company president acquired the big desk and the fancy office. The solution is to create a proposal that presents your idea as if it *were* reality. What you must present to your customer is *virtual* reality. The following are some things that the right presentation will say:

- Here's a picture showing what my product will look like when it's produced.

- Here's a prototype that shows how my product will work when it's produced.

- Here's a mock-up that shows how my product will be packaged when it's produced.

- Here's a drawing showing how stores will display this product when it's produced.

- Here are estimates showing how much profit you will make when it's produced.

I could easily add a dozen more entries to this list, but I'm sure you get the point. Because your customer buys reality, you need to sell reality. To the extent that your presentation can create that illusion, it will be your principal tool in closing the deal.

"They Stopped Laughing When I Took Out My Little List"

Here's a tip for writing your presentation. In teaching my college course, I noticed a phenomenon that actually made me laugh out loud when I saw it for the first time. Now I just expect it. I can stand in front of the class and talk nonstop for 45 minutes and just get glassy-eyed stares. I can be revealing the wisdom of the ages or just giving a weather report. No matter what, the unblinking looks remain. However, as soon as I say something like, "There are five steps in developing an idea" or "Here are the four important things to remember when making a presentation," all notebooks are thrown open, and everybody reaches for their pens.

So my tip of the day is this: *Everybody Loves Lists!* Smart people love lists and so do foolish people—and rich people and poor people and big people and little people. Every person who can read loves lists for the following five reasons:

1. Lists are easy to read.

2. Lists seem efficient.

3. Lists seem factual.

4. Lists seem to the point.

5. Lists seem conclusive.

When I say *lists*, I should make it clear that I'm including all the short-hand ways to make a point. That includes pie charts, bar charts, horizontal graphs, vertical graphs, and any other visual aid you can think of. *USA TODAY* has been living off the phenomenon for years. List lovers are everywhere. They're indistinguishable by sex, class, color, financial status, or birth sign. All of us are suckers for lists. So when you're structuring your own presentation, it's easy to organize it with headings like these:

- Five Reasons Consumers Will Buy This New Product.

- Three Reasons This New Product Will Decrease Your Production Costs.

- Eight Advantages This New Product Has over Existing Competition.

The briefer your proposal, the better the chances it will be read. And if they read nothing else, they'll almost definitely look at your charts and lists.

The Cost of the Written Proposal

The basis of your proposal is an intelligent explanation of what the product is, how it works, whom it benefits, and what's in it for the prospective licensee. If practical, your written presentation should also offer suggestions as to how your product should be produced, where it should be produced, what it will cost, what it can be sold for, where it should be sold, how it should be packaged, and how it should be displayed. Anything you can answer with authority should be addressed in your report. If you're just guessing about something, leave it out. And whatever you do, don't lie. You're dealing with professionals who will almost certainly uncover the fib, and that will cast a pall over everything else you've written.

The proposal will be much better if it includes attractive, full-color drawings or professionally taken color photographs to support the written exposition. Also, I cannot think of a circumstance when a presentation would not be improved by the inclusion of a prototype. Often, in fact, a prototype is mandatory. Most times a new product concept is viewed as merely an idea until it's reduced to practice. For instance, if your invention promises to operate in a certain manner or promises to achieve certain results, as the inventor you have the obligation to prove your claims. What better way than with a working prototype? If you want to be thought of as an inventor and if you want to receive the royalties that inventors sometimes do, then you have to *invent* the product. A claim on paper is not as compelling or believable as a prototype demonstrating that your idea delivers exactly as promised.

If you can put something three-dimensional in the prospect's hands, you will turn a dream into something practical. It turns into virtual reality. Something that a manufacturer can hold in her hands turns a dream into salable goods.

Out of the Twilight Zone and into the Profit Zone

I only create consumer products, so everything I license ultimately ends up in a retail store. As part of my presentation, I always include a mock-up package, complete with illustrations and copy. When I put it into the licensee's hands, we stop dealing in dreams and start dealing with visible reality. I even provide drawings to show how my product will look when it's displayed in a store, and the licensee *always* lingers over that image. We're now on the customer's turf: packaging, displays, merchandising. Through the strength of my visual presentation, I've managed to move my product concept out of the Twilight Zone and into the marketplace where a meeting of the minds can take place. The signing of the license is now a mere formality. Yes, that's a slight exaggeration, but I know you take my point.

If your product idea is truly inferior, no fancy presentation is going to help. And if your idea's truly brilliant, you may actually be able to sell it from a cocktail napkin. But most of our ideas fall someplace in between, and we need all the help we can get. We need the slickest presentation we can make for the following reasons:

1. *A Slick-Looking Presentation Shows Respect for the Idea.* A crudely prepared presentation implies a lack of confidence in the idea and exposes you as an amateur. A presentation that looks special makes the idea seem special. As I said before, even the most sophisticated executives will pause longer over an attractive presentation than they will over a crude one.

2. *A Slick Presentation Makes You Look Good.* When executives see a professional presentation, they are going to pay a lot more attention to what you're telling them. After looking at the presentation, they'll look at you with greater respect.

3. *A Slick Presentation Removes Uncertainty.* Businesspeople hate uncertainty. It makes them nervous and it makes them hesitant. A slick presentation chases away fears and doubts and makes the executive more interested in moving forward; it edges the new product proposal a bit closer to the "sure thing" side.

4. *A Slick Presentation Will Stand on Its Own Merits.* Your visual presentation perhaps will be shown to others in your absence, so it must work by itself without you being there to say things like, "And this little bump is supposed to mean ... "

This Is the Time to Take Your Best Shot

You'll probably have to spend some money here, but it's well worth it. The written portion of your proposal should be professionally typed or produced on a word processor. Find an artist to make the drawings, use a professional photographer, and if you can't do it yourself, get a model maker to do the prototype. A possible source of help would be a local inventors' club; another might be the head of the industrial design department at a local art college. This person can put you in touch with a talented student. It's good experience for the student and obviously beneficial for you. He's entitled to be paid, of course, but it'll cost far less than if done by a professional design firm.

You may even want to consider a videotape presentation. It can be quite impressive if it's applicable to your product. Inventors send me tapes all the time—and they're quite effective in presenting the idea. This method of showing concepts has become so prevalent in the toy field that virtually every marketing executive has a VCR set up in his or her office. We are in a video age, and if you can make your idea come to life on a video screen, you have a powerful sales tool. What could be a more effective way to turn a dream into virtual reality? One note of caution: If you do use the services of a professional artist or model maker, it would be prudent to have them sign a simple nondisclosure form before you turn over your material. Specialists like these are accustomed to signing these forms, so you can expect them to do it as a matter of course. A sample for your use is in the Appendix.

One reason franchises are so popular is that they remove so much decision making from the prospect. The franchiser's proposal spells everything out in infinite detail. The presumption is that by simply following directions, the franchisee will make a fortune. Every question is answered; every doubt is calmed. Your presentation, in its way, should attempt to do the same thing. If you've answered all the questions about your product's benefits and costs and if you've shown what profit can be made from it, all that's left for the licensee to do is nod in agreement. And nothing would make that person happier.

So make sure you have a first-class presentation. If you still can't sell your idea, at least you'll be comforted that you gave it your best shot. And if you do sell it, the money you spent on the presentation will have been more than worth your investment.

PRIORITIZING YOUR PROSPECT LIST MAY BE THE DIFFERENCE BETWEEN A BIG AND A MEDIOCRE SUCCESS

I hate to have to tell you this, but it's almost impossible to sell a new product idea through an unsolicited mailing to someone who doesn't know you. I've tried more than a dozen times over the years and have never been successful, and I don't know anyone else who has been successful either. If your invention is compelling to a manufacturer and if you can convey that urgency in a letter, I suppose you can achieve a sale by mail. Most of our ideas, unfortunately, aren't that powerful, and they have to be sold in person.

I'm always meeting people who want to get into the mail-order business. It sounds so great. Place a little ad for a weight-reducing elixir in *USA TODAY* and get bundles of checks in the mail. Many inventors feel the same way. Get a patent, send form letters out to a bunch of manufacturers, and sit back and sort through the contract proposals. Have you ever met anyone who got rich through little mail-order schemes? And have you ever met anyone who received money by sending out an unsolicited product proposal? If you have, then you and I don't travel with the same kind of people.

I can't stress enough the importance of personal presentations versus mail solicitations. Invention marketing companies, for a fee, promise to introduce your new product idea to "hundreds of industry leaders." All they do is make a mailing from bought lists. Big deal. Company executives accustomed to getting these mailings frequently don't even open the envelopes.

I have lately been browsing a few of the inventors' newsgroups and chat rooms on the Internet, and although the level of intelligent discourse is sometimes impressive, the subject matter is sadly revealing. Someone will post a question about the patenting process, and like flies to honey, immediately a half-dozen patent lawyers or semipro patent appliers join in. "Should I get a PPA?" "What about an NDA?" "Can someone clarify the first-to-invent laws here in America?" "How about a utility patent, or should I get a design patent?" "And what about the PCT?" "And the PTO?" And on and on. It's not that the questions aren't sincere or that the responses aren't often informed; it's just that so much of it is so screamingly repetitive and totally beside the point.

Assuming, of course, that the point is to make money by licensing ideas.

So much attention is paid to process that no one seems to have the time or inclination to ask about the *business* of inventing. No one asks, "How should I prepare my presentation?" "How do I handle the not-invented-here syndrome?" "Do I need to worry about store displays?" "How do I line up appointments at a trade show?" How do I get a professional appraisal of my idea?" "What are price points?" "Discounts?" "Markups?" Nobody asks these things; they only ask about patents and for advice about how to prevent someone from stealing their idea.

It's not surprising that so few patented products are ever commercialized. Many, of course, are simply not worth a licensor's time, but more often the inventor simply hasn't a clue about what to do next. Inventing they know; patenting they know; everything else is a dark void.

Frankly, when I wrote the original version of this book, I had no idea how lacking in information or interest other inventors were in the *business* of marketing their ideas. That's why the invention marketing companies continue to grow and prosper. They broadcast their siren song: "Mr. (or Ms.) inventor, you just keep dreaming up those wonderfully brilliant ideas and leave everything else to us," while they quietly and efficiently remove $10,000 or $15,000 from your wallet or purse. It doesn't have to be that way—you *can* do it yourself, and you may find that you actually like it. All you need to know is how to go about it, which this book intends to carefully outline.

> The cowards won't start and the weak will die along the trail.
> —Kit Carson, referring to the great trek westward

Yesterday I had a meeting in New York to present a new product idea to a major textile company. Including getting to the station, waiting for the train, getting to my appointment in uptown Manhattan, and returning home, my round-trip travel time was approximately five hours, and the cost was about $100. All for a 40-minute meeting.

This company has licensed many other products from me, and we have a continuing relationship. I could perhaps have sent my prototype and presentation by a package delivery service, but that possibility frankly never entered my mind. I know that by making the presentation in person, my chances of selling it were light-years better than they would be doing it by mail. I was able to pitch my idea, answer questions, head off objections, and leave with a clear understanding of the status of my proposal and what would happen next. My contact, who loved the product, is having a new product meeting in four days and will call me when it's over. Before the end of the day, I had sent a fax (paper trail) confirming the results of our meeting and restating when I expected to hear from the company.

This is how you get the order. If you want to do business, you have to enter the marketplace, grab the prospect by the lapels, look that person in the eye, and *sell* him on your new idea. There are no substitutes for face-to-face selling, so you might as well get ready for it. You don't want to be one of Kit Carson's nonstarters.

If You Can See Them, You Can Sell Them

When you make up your prospect list, always keep geographical proximity in mind. Obviously, someone from Chicago has an advantage over someone from Fargo. But who said life is fair? If you do live in a small town where there aren't many companies, you should be aware of this when you're developing your ideas. Take an inventory of the companies you can reach and focus your thinking on products for them. It's a strategy that will definitely pay off.

Developing the actual prospect list is easy. For the camping utensil idea, clicking on to some Internet sites or making a few visits to sporting goods stores and searching through catalogs and outdoor magazines should give you all the names you need. If that doesn't produce enough, looking through the *Thomas Register* online at your local library should provide plenty more. This thick 19-volume set lists by category more companies than you could believe existed in one country. There are plenty of other reference books as well; just ask the reference librarian.

Also, it would be worthwhile to ask the proprietor of a sporting goods store where and when the industry trade show is held. This is where all the sporting goods executives meet, and you could greatly expand access to your prospects by attending it. You can find the location and dates for just about any show you're interested in through a simple Internet search. Web site addresses for online directories are in the Appendix. You can almost always register as a guest to enter the exhibit hall. Sometimes you'll have to pay a fee, but most times you won't. In either case, you *must* have a business card.

God Bless Trade Shows

Trade shows are heaven for me. I go to housewares shows, gift shows, toy shows, electronic shows, textile shows, and general merchandise shows. I've been to shows in England, Hong Kong, France, Taiwan, Ireland, Italy, Japan, and the Philippines. If I'm currently interested in developing or selling some giftware items, for instance, I'll go to the New York Gift Show, which is the largest in the world. Here I find literally thousands of companies and their executives. I browse, I talk, I make contacts, I survey the market, I make appointments, I find ideas, and I make presentations.

If I have a product idea to present, on the first day or two I slowly walk through the entire exhibit hall, jotting down the names and booth numbers of every firm that's a prospect for my new product idea. I put them in order, from the most likely to the least likely. On the first few days of a show, industry representatives are usually too busy with customers to talk to you. By the third day, however, they're happy to have the company. I visit as many of the more likely prospects that I can. I introduce myself to the president or the representative in charge of new products and make a date to see that person the next day or later that afternoon, depending on our mutual schedules.

Whatever industry you're interested in undoubtedly has a trade show, and I'm sure you'll find it profitable to visit it. When I first went into this business, I was nervous and would take my ideas for new products to small companies. I thought I'd be treated better and that they would welcome new ideas more readily than large ones. This was a big error; I wasted some terrific ideas, and my royalties were a pittance compared with what they would have been with large companies. It's a lot harder to get in front of the decision maker in a big company, but it's not impossible—and it's well worth the effort. I'll talk more about this in the next chapter.

The obvious reason is that little companies are little because they don't sell many products, and big companies are big because they do. And big companies need fresh ideas more than little guys because they have larger engines to stoke and more money to spend. And they'll treat you just fine. So don't be hesitant like me and put the big companies at the top of your list. The little guys will always be there if you need them. Any successful salesperson will give you the same advice. If you were selling hammers, you wouldn't be calling on hardware stores if you could be spending the same time with the hammer buyer for Sears Roebuck.

DO YOU NEED AN AGENT? HOW TO FIND ONE AND HOW TO GET ONE

Now that you've evaluated your list of prospects, this is a perfect time to say a few words about agents. Should you contact manufacturers yourself or have someone do it for you? Benjamin Disraeli once said, "It is well known what an agent is: He is a man who bamboozles one party and plunders the other." Although agents are often the butt of jokes, in my experience they've been decent and helpful. Under certain circumstances, agents accomplished more for me than I could accomplish on my own, and I was happy to pay them their commission.

Who Are They?

My definition of an agent is simply someone who has access to decision makers, who will represent my interests to them in a professional manner, and whose remuneration is a percentage of the money earned by this effort. The key is that payment is in the form of commissions—not advances, not fees, and not retainers. Invention marketing companies may claim that they'll act as your agent, but they're not true agents by this definition. They won't do a thing without payment in advance, and their claim of access to decision makers is laughable.

I'm a professional licensing agent. I'm not in the research or patent business, so I have no reason to issue phony praise for your product idea in order to sell you other services. If I accept your idea and offer my services, I pay my own way and am rewarded only if I'm successful in landing a licensing deal for you. Invention submission companies, on the other hand, are fee-driven operations with little interest in whether the idea is licensed or not. They charge for so-called research, they charge for patent searches, they charge for patent applications, and they charge to supposedly submit your product to industry. Their standard spiel is to ask for a small royalty share if they're successful, but they would faint dead away if it ever actually happened. It's simply a ruse to enable them to say that they're charging you "only," say, $5,000 for their services because they expect to be earning most of their money from their royalty share. If they weren't doing so much financial harm to so many poor souls, their pitch would be laughable in its transparency.

True professional licensing agents operate in fields where it's difficult for inventors to meet directly with the prospective licensee. For instance, for the books I write, I use a literary agent because publishers prefer to deal with professional agents rather than with individual writers. Graphic artists also often use agents to offer their services to companies. This is also true in the toy and game fields, particularly with the larger toy companies who will deal only with professional agents whom they know and trust to not waste their time.

I also know of a few professional licensing agents who work in the giftware and home furnishings industries, but in most other fields, they simply don't exist. Because by definition the service that the agent provides is access to decision makers, they can operate only in areas where that service has value. For instance, if what you've developed is an industrial product geared to paint manufacturers, chances are you won't find an agent operating in that industry. Paint manufacturers don't have many inventors knocking on their doors, so if you are such an inventor, access to the manufacturer is easy. On the other hand, if you have a game idea to submit to someone like the Milton Bradley Company, you won't find such a warm reception. I don't mean to suggest that the folks at Milton Bradley are anything but decent and well meaning, but they are *inundated* with hundreds and hundreds of game ideas and they've long ago determined that it's in their best interest to look only at game ideas submitted by professional toy agents.

If your product is in an industry where agents work, you may feel the need to use one, but chances are that the agent won't feel a similar need for you. Most licensing agents I know turn down many more clients than they accept. If they don't think your product has merit, which is usually the case, they'll politely tell you so. By being selective, their value (and their access) to the decision makers is enhanced. The executive at a game company, for example, will see a professional licensing agent because it is an efficient use of her time. The agent is going to present a prescreened, small selection of games, each of which is worth consideration. The manufacturer doesn't know you or your work, so without a connection, it would be extremely difficult to get your own appointment.

On the other hand, if your invention is a terrific new attachment for an automatic metal-spinning machine, the manufacturer will probably tell you to come right over because not too many people have terrific new ideas for metal-spinning machines. Agents usually don't exist in these industries, but if they did, you wouldn't need them. As a rule, if you can get to the decision maker yourself, don't use an agent. You'll do a better job because you know more about your product than anyone else, and no agent can match your enthusiasm. On the other hand, if you can't get to the decision maker, I suggest you make an agent out of anyone who can.

If you do try to get an agent and are turned down, don't be discouraged or take it personally. All agents can tell you horror tales of products they rejected that someone else later turned into blockbusters. The interesting thing about this business is that *nobody* knows for sure what constitutes a winning product. They'll never admit it, but successful people simply happen to guess right more often than they guess wrong. Remember the three stages of an idea. First, they'll all say it's impossible. Next, they'll say it's possible but not worth doing. Finally, they'll say that they knew it was a good idea all along.

If you want to find out if agents exist in your field, you can do this by checking in your industry's "who makes it" directory. In the back, after the manufacturer lists, there's usually a section for salespeople and other service providers. If there are agents operating in your industry, that's where they'll be listed. For details, ask for the *Standard Rate* and *Data Service* at your local library.

What Do They Want from You?

If you decide to use a licensing agent and you can find one and the agent agrees to represent you, you'll have to sign a contract. I've seen some that are just one page and some that are six or more pages in length, but they all essentially have the same few basic points:

1. Commission arrangement.

2. Exclusivity:

   A. By area.

   B. By time.

3. Method of payment.

4. Cancellation procedure and rights.

Commission arrangements usually range from 10% to 50%, depending upon the agent's level of contribution and what he has to invest in the idea to bring it up to a professional level. Agents are quite perceptive and usually know what will sell. If they redesign your idea, make prototypes, and create professional presentational material, then their royalty share will naturally be high. Often their financial investment in the product is larger than the inventor's. For reference, I've included my own agent agreement in the Appendix.

Every legitimate licensing company will expect to represent you on an exclusive basis. Their territory is often the world. Usually, they want this exclusivity for a year or two; then they automatically renew it year to year thereafter unless either party cancels. This means that regardless of when or where in the world your product is sold, they get their percentage. If you happen to mention your product idea to a woman standing in front of you in a supermarket line in Bucharest who just happens to own a company that is looking for exactly that product and you license it to her, then your agent in New York still gets the same commission.

In your agreement, you'll give the agent the authority to negotiate on your behalf, although you retain final say-so. If your product is licensed, the royalty check goes to your agent. After the check has been deposited and has cleared, the agent will send you his own check, from which the commission has been deducted. Naturally, the agent will also send along the sales reports from the licensee.

How Do You and Your Agent Part Friends?

Your agent is required to keep you informed as to what companies have been called on and the results of the call. If you decide to fire the agent, he is still entitled to any royalties that may be earned from any of these companies, even if it happens years later.

Client-agent relations in most fields are volatile and often don't last long. The client thinks the agent isn't looking after her interests, and the agent thinks the account is a waste of time. This is the nature of these relationships, so it's not such a tragedy if you let your agent go at the end of the contract. Besides, an agent who hasn't sold your product by then will have no cause to object. Clients are so easy to find (particularly ones that generate no money) that the agent will hardly know you're gone. Just keep your parting on a professional level because business is business. Besides, you may need that agent again someday.

Use Special Agents Whenever You Can:
It's the American Way

Because what you're after is access to the decision maker at the company you want to connect with, that can come from many different sources. Lots of folks are not professional licensing agents but for one reason or another have the connection you're seeking. Perhaps you and the company executive have a mutual friend or share the same lawyer or accountant or insurance agent. Ask around, and don't be shy. It's just another form of good old-fashioned networking. I spend a great deal of time in Asia working on products for clients. There, everybody's second occupation is being an agent. Introductions are bought and sold all the time. They provide the grease that turns the wheels of commerce. It's the same in Washington. Arranging access to decision makers has made millionaires out of more than one former senator or retired general.

GETTING THE APPOINTMENT: THE MAGIC
TWO-MINUTE TELEPHONE CONVERSATION

Depending on your disposition, we're now at the point you've been waiting for or dreading. This is where the selling begins. In just the briefest of telephone calls you have to convince some tough, high-powered executive to make time to see you. Will she laugh? Slam the phone down? Tell you to go fly a kite?

You can relax. In most industries, getting an appointment is easy. You just need to stretch the truth a little and have a little nerve. After you do it once, it's fun. Just follow these simple, foolproof instructions.

Four Things to Remember When You Call for an Appointment

1. *You Are Not an Amateur!* Companies are disinclined to deal with nonprofessionals. Rarely do amateurs have anything worth showing, and amateurs are much more wary than professionals about having their ideas stolen. Having learned from experience, companies don't want to be sued by trigger-happy neophytes who think that everybody who looks at them intently is out to plunder their ideas. Professionals behave more rationally, often have more reasoned proposals, and are therefore automatically made welcome. So consider yourself at war—and the first casualty of war is the truth. You are *not* going to call the Green Velvet Mower Company and announce that you're actually a computer programmer by trade but you have a cute idea for a lawn mower. By innuendo, you're going to present yourself as a successful professional. I'll tell you in a moment what I mean.

2. *Go for the President.* You should usually try to reach the president of the company. You can't always do it, but you'll be successful more often than you think. If you can't get to the president, try for the number two person. If you're calling a giant company such as Johnson & Johnson, you probably wouldn't get through to the president, and it's unlikely that you would want him anyway; however, there is someone in charge of the division that interests you, and that's the person you want to target. Whatever happens, don't put yourself into the hands of some third-level assistant who promises to show your presentation to the boss (who is away on vacation right now). Nothing good can come from such a contact. Decision makers are deceptively accessible. Call one day and get the name. Call the next day and just tell the operator to connect you to good old Bill Duffy. It's really quite simple.

3. *Keep It Brief.* When you get Bill Duffy on the phone, keep your conversation as short as possible without being rude. Remember that all you're calling for is an appointment. When you get it, say thank

you and hang up. If you wind up chatting with Duffy on the telephone, you're going to say too much, and your appointment may get canceled right before your eyes.

Good old Bill is going to find out that you're actually a full-time computer programmer and that this is your first new product idea. Worst of all, you might get nervous and say too much. Before you know it, Duffy will decide he heard enough and that a meeting isn't really necessary. "Just put some stuff in the mail to me," he'll say, "and we'll let you know." As someone named Olin Miller once said, "When a person says he'll let you know—you know." Consider yourself a commando. Understand your mission (an appointment), make the raid (the call), and get out as soon as your mission's accomplished. If you hang around to see your handiwork, you're a goner.

4. *Niceness Pays.* Always be pleasant and friendly to receptionists, secretaries, and aides. They're often the keys to executive access. If you come across on the phone as a friendly person, they'll be inclined to help. If you're nasty, they may find a way to get even. It often happens that you'll be in some executive's office, showing her a new

Be as Brief as Possible When You Call for an Appointment

product concept, and she'll invite her assistant in. "Shirley," she says, "you just bought a new car. What do you think of this doo-hickey?" For your sake, I hope you were nice to Shirley.

Here is the script for a short, one-act play that I've acted in dozens and dozens of times. It's called *The Appointment:*

The Cast:

HELEN FURY	HECTOR GREENBOTTON
receptionist executive	vice president
MAX DUBOIS	BUSTER BALDWIN
President, Global	Hector Greenbottom's
Amalgamated Things, Ltd.	assistant
DONNA MADONNA	THOMAS ALVA REDDISON
Dubois's secretary	handsome young inventor

FURY: Hello, Global. Can you hold? (*Good. A busy receptionist doesn't have time for questions.*)

FURY: Hello. Sorry to keep you waiting.

REDDISON: That's okay. Let me have Max Dubois, please.

             (*Reddison will have called a few days earlier to get the president's name. The way he's made the request implies he and Dubois are old friends. If Ms. Fury asks any questions, Reddison just says he has something for Dubois and asks again to be put through.*)

FURY: Ringing ...

MADONNA: Hello. Mr. Dubois's office.

REDDISON: Hi. My name's Tom Reddison. We create new products for companies and have come up with something that I know Max will want to see—so if you could put me through, I'd appreciate it. (*Warm, friendly voice. If Reddison were an insurance person, he would need to use a different approach, but because Dubois is going to be curious, the direct approach is best. Note that Reddison says we, never* I.)

MADONNA: ... moment.

DUBOIS:	Dubois here.
REDDISON:	Hi, Mr. Dubois. Thanks for taking my call. I know you're busy, so I'll be brief. We design new products for companies, and we've hit upon something that I think can be very special for Global Amalgamated. If I could have a few minutes to show it to you, I promise you won't be disappointed. (*This a perfect 15-second presentation. Reddison has thanked Dubois, flattered him by acknowledging how busy he is, stated the purpose of the call, piqued his interest, and asked for an appointment. The use of* we *suggests one company contacting another.*)

At this point, Dubois will say one of three things:

1. Okay, when do you want to come in?

2. Sounds interesting. What have you got?

3. Oh, you want Hector Greenbottom. He handles all the new products.

If Dubois's answer is number 1 (When do you want to come in?), Reddison makes the appointment and hangs up promptly after giving the assistant his telephone number in case he has to cancel. If nobody knew how to reach Reddison, he might drive from Pittsburgh to Cleveland only to find that Dubois was out sick. It happens.

If Dubois's answer is number 2 (What have you got?), Reddison answers as follows:

REDDISON:	I really can't do it justice by trying to describe it over the telephone, but when you see the drawings [or model], I know you'll be excited. It fits with what you're doing, and I promise I won't be wasting your time. (*Dubois agrees to the appointment.*)

If Dubois's answer is number 3 (Oh, you want Hector Greenbottom), Reddison asks to be switched.

BALDWIN:	Mr. Greenbottom's office. (*Baldwin is a screener. He's authorized to say no, but he can't say yes. Avoid him at all costs. Baldwin learned a long time ago that you can get into trouble by saying yes, but you almost never get into trouble by saying no.*)

REDDISON: Hi. My name's Tom Reddison. We create new products for companies, and we've recently developed something for Global that Max wanted us to show to Mr. Greenbottom. So if you could put me through, I'd appreciate it. (*Reddison speaks to Greenbottom, repeats his pitch, fends off a request to describe the product, and makes the appointment.*)

Except for any waiting time, this whole exercise should take only two or three minutes, and you almost certainly will get an appointment. Never, ever describe your idea on the telephone. It takes the mystery out of your visit, and you run the risk that the person on the phone will lose interest. If you simply explain that a telephone description won't do it justice and that the person is really going to like the product, you'll almost always get the appointment. Sometimes, out of curiosity, the person will say something like "Okay, but can't you just give me a hint?" Just give a warm chuckle and offer reassurance that your visit will not be disappointing. I've almost never been denied an appointment once I had the right party on the telephone. Every company executive wants to learn about a new idea that may benefit the company. These people didn't get where they are by being foolish. They just want to be sure they won't waste time with some amateur who doesn't know anything about the business.

I'd like to tell you a little story that doesn't necessarily put me in a favorable light but illustrates these points better than any way I can think of.

I live in Philadelphia, and there's a little company here that makes a few products for the big stationery supply companies such as Staples, Office Max, and Office Depot. I had a product idea that I thought might interest this company, and although its competition was bigger, I decided to give it the first shot, if only because it was so close to home. Lazy slug that I am, it seemed like a sensible move.

I did what I always do—I called the company and told the operator that I had to send a letter to the president and needed his name. She readily supplied it, and a few days later I called again asking for "George Allen" (not his real name). A few clicks and the operator was back again. "May I ask what this is in reference to?" "Sure," I said, "we're a product development company and have come up with something that I know he'll be very interested in seeing." Again a few clicks and then: "I'm sorry, Mr. Allen has stepped away from his desk. Can you call back later?" This was obviously a stall—but I'm a big boy.

Over the next few weeks I called Mr. Allen more times than I can remember, always to be told that he had stepped away from his desk. It became a game. I put him on my speed dial and whenever I thought of it, I pushed the button, laughing to myself with the knowledge of what the response would be. Finally, I sent Mr. Allen the following fax:

Dear Mr. Allen:

Over the past few weeks I've called you more times than I can remember, only to be told each time that you were "away from your desk." If I were thinner skinned, I'd think you were avoiding me—but then I ask myself: "Why would he? We're an internationally known product development company, creating new product concepts for some of the most famous names in American merchandising. If I'm calling Mr. Allen (I tell myself) because I think we've come up with a new product idea that would interest him, why in the world would he avoid me?"

Because I could not come up with a reasonable answer, I must assume that it has actually been a case of miraculous bad timing—and that each time I called, you really were away from your desk. That being so, I'm faxing now to request a brief (about 10- to 15-minute) appointment. If you'll fax me with two or three available times, I'll pick one and confirm back.

(Signed) Harvey Reese

The very next day I had a message on my voice mail from a "Ms. Jane Smith" (another fictitious name), identifying herself as the director of marketing for the company. This prompted the following fax:

Dear Mr. Allen:

I had a call on my voice mail from Jane Smith, presumably in response to my previous fax to you. Please extend my apologies to Ms. Smith (whom I don't know), but I won't be calling her back. If I wanted an appointment with her in the first place, that's whom I would have called—and frankly I find it disrespectful of you to turn me over in this manner.

However, apparently you'll be pleased to know that I won't bother you any more and will simply move on. Fortunately, there's no scarcity of companies that know us and are always anxious to see whatever new product ideas we've developed for them.

And that's the end of the story. I never did meet Mr. Allen and moved on to place the item elsewhere, just as I told him I would. This was no great triumph for me, but it serves to illustrate two important points:

1. *This is a very rare event.* Over the years I've set up hundreds of appointments for myself, and I can literally count on the fingers of one hand the number of times I didn't sit down in front of the person I wanted to see. The lesson is to *not* be afraid to ask for an appointment. If you do it right, you'll almost always succeed. Mr. Allen turned out to be the exception to the rule. Nearly every time, the person will agree to see your new product idea. All she wants is to be convinced that you're not weird and that she won't be wasting her time. There are tricks to getting the appointment, which I've already discussed. There is *no* substitute for a personal visit, and if you can't or won't do it, then you should try to get someone like me to do it for you. It's what I enjoy.

2. *You must meet with the decision maker.* I know from long experience that it's a waste of time to meet with anyone other than the person who can say yes. Mr. Allen's company was a small one—and no matter how many titles are given out, in a small company it's the owner who makes the decisions. I could have met with Jane Smith, and she'd have dutifully taken the information, but I know I'd be wasting my time. The Jane Smiths of the world are gatekeepers. They have the authority to say no but are never given the authority to say yes. In small companies if you don't see the guy who signs the checks, you're wasting your time. After all, the company is paying people like Ms. Smith to dream up clever product ideas, so why should they pave the way to make a star of an outsider?

I moved on and licensed the concept elsewhere, as I told Mr. Allen I would. It's his loss, not mine. There are far more companies looking for fresh new ideas than there are folks who can supply them. Just remember—don't be put off if you run into an unresponsive executive once in a while, and make sure you turn on your charms only to the person who can say yes.

Making a Mail Solicitation That's Slightly Better than Nothing

Even if it's not possible for you to make a personal sales visit and you therefore plan to mail the material, you should still call first. Your telephone conversation will be similar to the one we just reviewed, up to the point where you ask for an appointment. Instead, you'll tell Mr. Dubois that you have this terrific idea and you'd like to send him some information. Find a reason to chat with him for a moment or two so he'll remember you when your material arrives. Your cover letter can then remind Mr. Dubois of your telephone conversation, thank him for his courtesy, and state that the enclosed material is being sent at his suggestion. It's not as good as a personal visit, but it's much better than an unsolicited mailing.

And while we're on the subject of material, *never* send your original! With the best intentions, material may get lost or thrown out, or you may have to wait months to have it returned. In the meantime, you're dead in the water. For color work, find a color copier. The color reproduction will be almost as good as the original. And send the material to Mr. Dubois via an overnight delivery service or the Post Office's express mail service. Certainly, it costs a few dollars more, but it makes your presentation special.

**Once You've Sold a Company Your First Idea,
It'll Be Happy to See Your Next One**

Confessions of a Habitual Failure

As previously mentioned, personally, I've never ever made a successful new product submission by mail to a company that didn't already know me. Even though I've always called the prospect first and even though I've always chatted with the prospect on the telephone and even though I've always sent a first-class presentation, I have never been able to close a sale. I have tried *many* times, but the caliber of my ideas must be such that I'm unable to get people excited with a letter. If pitching your product by letter is the only way you can do it, I hope you have better luck (but don't say I didn't warn you).

On the other hand, the good news is that once you've sold one product to a company, it's far more inclined to buy a second and third from you regardless of how you transmit it. If the company knows you and trusts you, you'll find that the e-mail scanner is the best invention of the twenty-first century.

Even Grown-Ups Have Wish Lists

In the toy industry, where every year means a new battle in the marketplace, companies cater to established toy developers. It's fair to say that many smaller companies owe their very existence to independent product developers. They invite them to private showings of their line, they hold cocktail parties for them, and most importantly, they provide wish lists. These lists outline the areas in which the company is looking for toys for the following year, with entries such as "We need craft items, outdoor water toys, and electronic games." If you've created something worthwhile that happens to be on the list, you can tie a sketch to a brick and throw it through the president's window. The company will still be happy to buy it from you.

If the primary reason to conduct your business in person is that it greatly improves your chances of success, the secondary reason is that you get to know the company and the kind of products they're looking for—and they get to know you. You'll never acquire this kind of knowledge or establish this kind of relationship through the mail. When you know what the company wants, it helps to focus your thinking for future products. The company supplies the problem; you supply the solution, and before you know it, a valuable alliance has formed.

When you proceed to dream up a new product idea, it helps a great deal if you already have a company in mind to sell it to. You can create your first idea in a vacuum, but if you intend to do it over and over, it facilitates matters greatly to have a target company in mind. Once you've licensed your first product to a company, you'll probably be able to continue doing it. You can't develop relationships through the mail, so the sooner you start phoning for appointments, the better.

THE CONCEPT OF THE NONDISCLOSURE FORM: DO YOU SIGN THEIRS, OR DO THEY SIGN YOURS? AND DOES IT MATTER?

For the sake of our discussion, let's assume that the product you've created is an attachment for an automatic lathe. Manufacturers of industrial equipment don't get that many calls from people who have great ideas for them, so your telephone call will almost exactly follow our script, and you'll be invited to come right over.

On the other hand, if you've developed something like a new toy or kitchen gadget, you may run into a different reception. Large, high-profile consumer goods companies have people contacting them *all the time* with ideas. These companies devour new products at great speed and in big gulps, so they're always looking for something fresh and exciting. If you have next year's hot toy or an innovative automatic bread maker, they definitely want to hear from you. However, as the old saying goes, You have to kiss a lot of frogs before you meet a prince. The truth of the matter is that most new product submissions aren't worth the powder to blow them up. The ideas are stale, half-formed, or half-baked, or they have no relevancy to the company's business. Almost all of them are a waste of a company's time.

To make matters worse, many amateurs who submit these concepts have an unrealistic estimate of their idea's worth as well as a possessiveness that borders on paranoia. And even if the product submitted does have merit, there's still the possibility that the company is already working on it. The company then runs the risk that the inventor will cry foul and go running to an attorney. The defense that some large companies use is a product submission form, more commonly referred to as a nondisclosure agreement. The company says, "Hey, we'd *love* to see your new product idea, and we bet it's terrific! One little thing though; please sign this form so we know that you understand and accept the way we operate."

Read It and Weep

1. The company says to please not send your idea until you've given it all the legal protection you can because that's the only protection you can rely on. Patent it if you can, or at least copyright it. If you don't do either, proceed at your own risk; the company assumes no responsibility.

2. You are making the submission on your own initiative and by your own free will. Nothing has been promised to you. Further, you understand that it's possible the company may already be working on the same or a similar idea, which makes your submission redundant.

3. You understand and agree to the following conditions:

   A. The company does not agree to keep your idea confidential.

   B. The company does not promise to return your materials.

   C. The company doesn't promise to pay you anything for your idea.

The only recourse you have is whatever protection is provided by your patent or copyright. If your idea doesn't have either, you have no protection whatsoever.

And have a nice day.

The Mead Corporation is an enlightened company, anxious to see whatever new ideas you might wish to show the company. Nevertheless, if you make contact, you'll receive the following form letter from its legal department, which it has allowed me to reprint here:

Dear _____

Mead is always receptive to new ideas and designs, as can be seen from the enclosed write-up, "Mead Welcomes New Ideas," which also outlines the conditions under which new ideas and designs will be evaluated. If you agree with the conditions as set forth in the write-up, please sign and return the form as indicated, together with the idea or design you would like Mead to evaluate, to my attention at the above address.

Your idea or design will then be evaluated and, thereafter, we will relay to you the results of the evaluation.

Thank you for considering Mead in this matter. We look forward to hearing from you in the near future.

Very truly yours,

Included with the letter is the following:

MEAD WELCOMES NEW IDEAS

The Mead Corporation welcomes ideas or designs and will evaluate outside submissions if the idea or design relates to our business.

Of course, we are not interested in ideas or designs that we may already know about, or that are in the public domain. We have many projects on which we are now working and many inventions of our own which we have not yet put into operation.

To protect you, as well as ourselves, we must follow a definite procedure in handling such ideas and designs.

WHAT SUBJECTS ARE ACCEPTABLE

Your idea should encompass a new or improved process, machine, product, composition of matter, or design relative to some phase of Mead's business. In general, Mead's business interests include paper and related products, paperboard and related products, metals and minerals, interior furnishings, consumer and educational products, pulp and forest products and information technology.

DESIGN SUBMISSIONS

Frequently, designs and ideas for the use of certain designs are submitted to Mead. These designs are often in the form of drawings, sketches, photographs or the like. You should note that mere general suggestions are treated by the law as abstract ideas and only the specific embodiments or expressions are subject to protection by design patents or copyrights. Therefore, you must keep in mind that in submitting your design to Mead for evaluation, you agree that your submission is limited to the particular design(s) submitted and does not encompass whatever marketing or other generic concept might be associated with such design(s). By the way of example, if you submit a photograph of a scenic river and suggest its use on notebook covers, your submission is limited to the particular representation appearing in the photograph you submit, and would not extend to other natural scenes, whether or not such scenes include scenic rivers.

YOUR PROTECTION

The United States patent system protects inventors against unauthorized use of their inventions and the United States copyright system affords protection to authors against copying of their work. Therefore, for the protection of both you and Mead, you must understand that, in submitting and thus disclosing an idea or design to us, you rely upon your patent or copyright rights, present or future, for your protection against unauthorized use of your invention or design. You should fully protect your idea or design to your own satisfaction before submitting it for our consideration. You should bear in mind that in submitting your idea to Mead under a non-confidential relationship, you may well be jeopardizing or forfeiting patent rights in certain foreign countries in which you have not yet filed patent applications.

HOW TO SUBMIT YOUR IDEA

We prefer that you either apply for or obtain a patent or copyright your design before sending it to us. When this proves impossible, make a written record of your idea or design in duplicate. Sign and date both copies and have one or more persons to whom you have explained your idea or design sign and date them, too. One of the copies can then be submitted to us, in accordance with the conditions outlined herein. Keep the other copy. While such a written record does not, of itself, give patent or copyright protection, it would be of use in proving priority of invention, should the need arise. Further, this provides a record of what you disclose to us.

WHAT HAPPENS AFTER AN IDEA IS SUBMITTED

When we receive the completed form and your written idea or design, we will examine the form and subject matter. Any written idea or design which is not attached to a completed and signed form, or does not deal with the subject matter previously listed, will be returned to you.

If your idea or design is properly submitted and the subject matter is within our range of interest, we will review it and give you our evaluation of it.

And here, in conclusion, are the conditions you must agree to before sending your new product concept to Mead Corporation:

CONDITIONS OF SUBMISSION

1. All submissions or disclosures of ideas and designs are voluntary on the part of the submitter. No confidential relationship is to be established or implied from consideration of the submitted material.

2. With respect to unpatentable or uncopyrightable ideas and designs, Mead will consider them only with the understanding that the use to be made of such ideas and designs and the compensation, if any, to be paid for them are matters resting solely in the discretion of the company.

3. With respect to patentable or copyrightable idea and designs, including those covered by patents pending, patent applications and registered or unregistered copyrights, Mead will consider them only with the understanding that the submitter agrees to rely for his protection wholly on such rights as he may have under the patent and copyright laws of the United States. With respect to ideas and designs included in this subsection, Mead may want to negotiate with the submitter about acquiring rights therein.

4. It is understood that idea and design submissions or disclosures are limited to the specific embodiment or expression submitted and do not extend

Reading the Company's Nondisclosure Agreement

to any marketing concept or other generic concept which may be associated with said specific embodiment or expression.

5.   The foregoing conditions may not be modified or waived.

As forbidding as these documents may sound, the conditions are not unreasonable from the corporation's point of view. After all, it didn't come to you; you came to it. The bright side is that companies like Mead would *never* steal your idea, and if what you show them has merit, they'll be pleased to negotiate with you to get it. So don't let the wording alarm you. It's the company's game, and if you want to participate, you have to play by the company's rules.

If I call a new company for an appointment and the person I'm planning to see tells me that the company has a non-disclosure form, I tell him fine, just fax or e-mail it to me. I sign the form and send it right back, sometimes while we're both still on the telephone. Rarely do I bother to read it because I already know what it says. I know my rights, and I know I'm not the person the company's trying to avoid. It may be rash for me to suggest that you sign so easily, but you're really not giving away any legitimate rights, and these companies won't look at your idea unless you comply. They must protect themselves from cranks, and I know that description doesn't apply to you.

Some attorneys and well-meaning friends may suggest that you have your own nondisclosure form that you should get the company to sign before telling them about your idea. The company would promise to keep your information strictly confidential, they would promise to return your materials as soon as you requested them, and if no deal were made, the company would promise not to commercially profit by whatever they learned from you. These are all totally reasonable requests, except that few companies are willing to sign these things. You'll only spend your money needlessly and delay the process.

My area of knowledge is in consumer products, and I know from long experience that these companies don't sign nondisclosure agreements. However, if you have invented some highly technical and significant new process, material, or machinery, it's quite possible that the company you intend to demonstrate it to will sign such an agreement, and you might rightly back away if it won't. If what you have, however, is something like a new kitchen gadget, don't worry that the company you show it to is going to steal it. It's not good business, and it's cheaper for the company to simply make a deal with you if the idea is worthwhile.

In my own business, where I invite inventors to show me their unprotected ideas to evaluate and, hopefully, offer my assistance, I automatically provide a nondisclosure agreement to allay any fears. A copy is enclosed in the Appendix, and you're welcome to use it for your own purposes where applicable.

We have our appointment, we've signed the company's nondisclosure agreement, and our presentation is as crisp and professional looking as it could be. Now we move on to my favorite part of the C.R.A.S.H. program. We're ready to meet eyeball to eyeball with the person who's going to pay a fortune for our new product idea.

5

Meeting with the Prospect

How to Make a Successful Presentation

I don't like money,
but it quiets my nerves.

—*Joe Louis*

I'm one of the lucky people of the world because I love the work I do. I wear jeans almost every day, I shave about twice a week, and while everyone else is rushing to work, I'm enjoying my second cup of coffee. I work at home, at my own pace, and every month the royalty checks get delivered in the mail.

"Do what you like, do it the best you can, and the money will follow." I've heard that advice offered countless times over the years, usually by someone older to someone younger. It didn't make sense to me when I was a young man, just as it probably doesn't make sense to young people today; nevertheless, I'm living proof that it works. Until middle age, everything I did was for the purpose of making money. When bankruptcy caused me to change careers, I decided to just do what I like and to do it the very best way I knew how. And, sure enough, the money did follow.

And the part of work that I like the most is taking my latest product idea under my arm and going out to sell it. It's a special event because I put on a jacket and tie and I get to go someplace. And what could be more rewarding than getting someone excited by the product of your own creativity? It's the Rolls Royce of selling.

NO CALL IS A WASTED CALL

I don't make a sale on every call, not by a long shot, but each call is valuable, and I'm always glad I made it. If my product isn't right for the company I'm calling on, an executive there will often direct me to the company it *is* right for. "Go see Billy Kramer over at Kramer's Thingarama. This should be perfect for him. And be sure to tell him I sent you over!" And even if company representatives don't like my idea, they'll frequently offer suggestions that help improve it. Sometimes they point out fatal flaws in my idea. When that happens, I simply file the design away and move on to the next project. So no call is wasted, and if nothing else, I may have made a contact for my next idea.

I approach each call with the confidence that my prospects are going to love what I show them. Carnegie called it *the power of positive thinking.* I don't know why it works, but it does. It's an aura of success that all good salespeople have, and they don't even have the personal satisfaction of selling their own creation. Very few people can create an original, commercial idea, so you should carry yourself into the meeting with great pride. It'll reflect in your voice and in your actions and, most important, in the product you're presenting. Your accomplishment commands respect, and in most instances, you'll get it.

A SALESPERSON'S WORST NIGHTMARE (PART I)

I recently had an appointment with a company president whom I had never met before. I showed up for my appointment exactly on time, as is my habit, and was kept waiting in a tiny, windowless reception room for about 40 minutes. Finally I was ushered into the great woman's office. I wasn't greeted with an apology or even a hello. I wasn't even offered a seat. Instead, as I entered, I was abruptly informed that she was quite busy and could give me only 10 minutes. I should also mention that she knew I had driven more than two hours to get to her office.

I realized that there was no hope in saving the situation, so I told her to forget it. I told her politely but firmly that what I had was worth more than 10 minutes and perhaps one day I'd come back and we could start over. I must confess that the stunned silence was music to my ears.

Over the next several weeks she had her assistant call me on three different occasions to ask if I was ready to reschedule the appointment, and each time I replied that I was not. She obviously realized that she might have let something of value slip through her fingers. Just between us, it was actually not such a sensational product—I suspect her imagination did more for it than I could ever have done. I never did go back and eventually sold the product to her competitor. To this day she doesn't know what it is, and neither does she know what subsequent products I might have shown her if she had made me feel welcome. It's not a matter of being cocky. I simply felt I was prepared to offer her something of value and was therefore entitled to a degree of courtesy. When you're a successful product developer, you can afford to do this. There are so few of us and so many companies that need what we're selling.

A SALESPERSON'S WORST NIGHTMARE (PART II)

As a very young man, I had a job as a salesperson on the road. On my first day, on my very first call, I had an experience that will stay with me forever. My appointment was with a supermarket buyer, and sitting in the waiting room I was as nervous as any neophyte would be. Finally my turn came, and I was directed to his office. It was a typical buyer's office—small, airless, piled high with papers and samples and old wooden furniture. I felt as if I were visiting a cheap detective. And there he sat, waiting for me with hooded eyes, open collar, no smile, and no greeting. I said hello; he nodded. As soon as I was seated, he reached over, cranked an egg timer, and said, "Okay, kid, you've got three minutes."

To my credit, I didn't pass out. The ticking of the timer was like Big Ben on speed, each ticktock bringing me closer to my doom. Flustered, embarrassed, and rattled, I broke all records racing through my presentation (I must have done it in less than a minute), plunked down my catalog sheets, mumbled my thanks, and stumbled out of the office.

I'd like to tell you that he actually placed an order and that we ultimately became the best of friends. But he didn't and we didn't. What I did get out of the experience, however, was much more valuable. I vowed never again to allow myself to be treated this way by a buyer, and it never did happen again. That is, not until 30 years later when a company president gave me 10 minutes to tell my story. From 3 minutes to 10 minutes is not very much progress, when you allow for inflation.

Actually, little episodes like these are rarities. What's more typical is to be treated with hospitality and consideration by the people you call on. They are *hoping* that you're going to show them something special, so they'll greet you in an anticipatory mood. Executives love to see interesting new products. It's the lifeblood of their business. Other salespeople would be thrilled to have their prospects greet them in this frame of mind.

ESTABLISHING YOUR CREDENTIALS

The natural way to start any meeting is to tell something about yourself and why you're taking up this busy person's time. People have a preconceived image of an inventor that is not always flattering. The adjective *crackpot* regretfully leaps to mind and makes licensees a little uneasy. Let them see what a nice, bright, normal person you are and that you know what you're talking about. It'll put your prospects at ease and make them more receptive to your presentation.

When I call for an appointment, I never describe myself as an inventor because it tends to be off-putting. I call myself a product developer, which has a more businesslike tone. I want to be viewed as a businessperson whose business it is to sell profit-making ideas.

Allowing for obvious differences, male and female executives dress the same in similar circumstances. A female marketing executive for a Fortune 500 company projects the same sartorial image as her male counterpart. Therefore, either way, I'm able to dress similarly and can present myself as someone who fits into the environment. As best I can, I become like my prospect in dress and speech so that he or she will be more mentally prepared to hear my story. The signing of a license is usually the beginning of a relationship, so if you can give clues that it will be a pleasant one, you're that much ahead of the game.

Let's say you're sitting with Mr. Stewart Bunyon, president of the Universal Camping Equipment Company, preparing to show him your new idea for biodegradable, disposable eating utensils. You're obviously going to gloss over the fact that you're a computer programmer by profession. But you *are* going to stress that you're a veteran camper of 18 years, that you've camped out in 27 states and three provinces in Canada, and that you've been using Ajax Camping Equipment since way back when your dad used to take you out. Mr. Bunyon is obviously a camper himself and perhaps has visited some of the same camps. You will have established a foundation to build your relationship. Whatever your credentials are, this is the time to establish them. It's your audiovisual business card.

IT'S YOUR PITCH, SO PUT SOMETHING ON THE BALL

Although you're not exactly selling cemetery lots, you're a salesperson now, and any salesperson's first job is to establish rapport with his customer and earn some respect. You're going to want Mr. Bunyon ("Call me 'Stu'") to accept some of what you're about to say on blind faith, so it's important that he like and trust you. You had good background reasons for inventing *this* product and showing it to *this* company, and that information should be on the table from the very beginning. It is important for Stu to recognize your knowledge about camping because it adds legitimacy to your product idea.

I've met a number of typical inventors over the years and find that many are so wrapped up in the *process* of inventing that it becomes an end in itself. Most inventors are, first and foremost, problem solvers. The actual act of inventing and finding the solution is what keeps them going, and they're too timid or reluctant to invest time in selling their products. It's somehow beneath them. That may explain why such a large percentage of patented products are never commercialized. Many inventors hate the idea of going out to peddle their wares; they feel that should be somebody else's job.

If that was your attitude before reading this book, you'd better lose it now. If you can't go out and make an enthusiastic sales presentation for your own idea, who can? Seeing the commercial version of your product in a beautiful box on a retail shelf can also reflect the joy of creativity. And there is true pleasure in knowing that every time somebody buys one of your products, anywhere in the world, you're being paid for it. If you want to succeed, you must remember that you're not in the invention business. You're in the profit business. Anyone can get a good idea; almost everybody does from time to time, but what separates the professionals from the amateurs (and what makes some product developers very rich) is turning a dream into a commercial reality. Nothing wonderful is going to happen to you or your idea until you go out and sell it, and this is the time to begin.

THE TRUTH WILL SET YOU FREE

Now that you're about to swing into your presentation, you must be totally honest in your claims and representations. I know I encouraged you to gloss the truth a bit to get the appointment, but that was then and this is now. Righteousness has nothing to do with it; honesty is simply a matter of good business sense. So much of what the customer thinks of your idea has to do with what he thinks of you that only a fool would jeopardize the relationship with false statements. If there's any interest in your product and you make any untruthful or exaggerated claims, I promise you'll ultimately be found out, and when that happens, it may be an extremely expensive lesson.

SAY HELLO TO THE NEW PRODUCT DIRECTOR

"Frank, do you have a few minutes? I've got a guy in my office with a new product idea you oughta take a look at." Frank, it turns out, is the company's research and development director and, you must assume, commands Stu's respect. If you're automatically suspicious that Frank could be a deal breaker, your instincts may be correct. A new product director called into a boss's office to give an opinion about a new product idea from an outsider is not inclined to be your ally. It goes with the territory. Here's what I mean.

The Dreaded NIH Disease

I know a woman who's in charge of product research for a giftware company in Long Island, New York. For years I would bring her new product ideas, and for years she turned me down. She's a nice person and was unfailingly polite and interested when I made my presentations. But although she'd always ask that I leave my drawings and prototypes for further study, the ultimate answer would always be the same: "Sorry, but we decided it's too expensive [or too cheap or too big or too small]."

I knew the products I presented were as good as or better than what her company was selling, and giftware companies always need new products, so I was baffled by my repeated failure. One day, out of the blue, I finally realized the problem. The poor woman was afflicted with the dreaded NIH disease—"Not Invented Here"!

NIH is a widespread sickness that usually affects midlevel marketing people, designers, engineers, and anyone else charged with the responsibility of creating the company's new products. On the assumption that your success makes them look bad, helping you to fail is their pleasure. It's an insecurity-driven illness that can sometimes be contained by massive doses of compliments and outward manifestations of respect. Confrontational treatment usually results in severe relapses.

So with this in mind, you must be prepared to handle Frank with the assumption that he's similarly afflicted. The trick is to make him look good without making yourself look bad. Here's a snippet of the kind of conversation that you must avoid:

STU: So what do you think we can get for it?

YOU: I think ten dollars seems just about right because Consolidated sells their utensil set for seven ninety-five and ours has all these extra features. . . .

FRANK: So whatta you think it'll cost to produce?

YOU: I've looked at that very closely, and it shouldn't cost more than about three dollars, depending on whether you go with plastic or wood.

A smile creeps across Frank's face because now you're in big trouble. He can see that you didn't do your homework. You should know that in the camping business, a manufacturer needs at least a five-time multiple from cost to retail to make a profit. So if the cost to produce the item is $3.00, the company needs a retail of at least $15.00. And you've already said the optimum retail is $10.00. So now what? Were you correct at either end, or were you guessing? Do you *really* know that $10.00 is the best retail? Are you *really* sure about the cost figures? You get the idea.

The question of manufacturing costs versus selling price comes up at virtually every meeting, and I always duck it if I can. Let's rerun the scene:

STU: So what do you think we can get for it?

YOU: Well, you and Frank know pricing strategy far better than I do, but it's certainly worth more than Consolidated's set. I think consumers will gladly pay a bit more for our version. What do you think the right retail should be? Obviously *(turning to Frank)*, you know more about marketing than I do.

FRANK: Okay, first tell me what you think it'll cost to produce.

YOU: I have a rough cost breakdown here *(shows figures)*, but with your company's production skills, I know you'll do much better. The important thing is, even using my figures, you can see that there's ample room for profits. To be perfectly honest, Frank, I trust your judgment on this better than mine. It's obvious that you know what you're doing.

I then sit back and let Stu and Frank work it out between them. They do it all the time, and after a little give and take, they quickly come to a resolution, and the product has now become theirs. I don't look bad by saying I don't know, and I've shown respect for Frank's ability. I'll keep the conversation in this tone and hopefully will have Frank on my side before the meeting is over. Even if I really did have an opinion of the best retail and knew the precise production costs, I would rather involve the prospect if I could because it conditions the person to start thinking of the product as belonging to the company, not to me. And that, after all, is the whole purpose of this exercise.

FIVE SIMPLE TIPS, ALL WORTH REPEATING

Your written proposal will pretty much dictate the agenda of the meeting, so here are a few points to remember as you get started:

1. Whatever you do, *don't read your presentation.* Your customer will start to doze off, and your credibility will fly out the window.

2. Don't give out a copy of your written presentation *until you have fin-ished* your oral presentation. It will encourage your customer to read a page or two ahead of you instead of listening to you.

3. *Move along at a brisk pace.* Your audience of one (or several) catches on quickly, and if you drag things out, their minds will start to wander.

4. Don't unveil your prototype until the proper time. And when you bring it out, *place it in your prospect's hands,* implying that it's not yours anymore, it's the customer's.

5. If you need extension cords or batteries for a projector or recorder, bring them with you and *have everything set up before you begin.*

These basic rules will help you give a superior presentation. Many of your customers have spent so much of their lives in worthless meetings that they appreciate the occasional good ones. It will reflect well on you and your new product.

Rule 1: Never Read Your Presentation

GATHER AROUND, BOYS AND GIRLS: IT'S STORY TIME

When I proceed with my presentation, I do it as if I were telling a story to a child. Everyone, no matter how mature or sophisticated, likes a good story with a happy ending. Maybe I don't actually start with "Once upon a time," but the implication is there. Top salespeople will tell you that one of the most effective ways to make a point is to tell a personal story that has a bearing on it. Have you ever met anyone who wouldn't pause to hear a good story? My stories all have to do with how silly old me just happened to dream up this terrific new product that brought us together. Shucks, anyone could do it:

> One day, I stopped to watch those guys putting up the new IBM building down on Spruce Street when, out of the corner of my eye, I happened to notice …

> I promised my wife I'd pick up a new hair dryer, so I was in Kmart over at the Market Village Mall, and I wandered over to the automotive department and was amazed to see …

> I popped into the new computer store they just opened next to Ellworth Drugs on Pine Street when I saw this big crowd gathered around a display. I went over to check it out and couldn't help laughing when I found …

If I can get my prospects to agree that a problem actually exists in the first place, I have a good chance of convincing them that I've solved it. And like all good stories, mine has a happy ending. It's about how this company will live happily ever after on the big profits it'll make with my swell new product.

At some point in your presentation, you have to be prepared to show why the prospect should manufacture and distribute your product. It's not enough just to say that the company will make money with it. That's the obvious reason for the meeting. If you'll recall, at a certain point earlier in the C.R.A.S.H. process you had to evaluate your idea carefully before deciding whether or not to proceed with its development. The same factors that swayed you then can be used to sway your prospect now, except now you state them as conclusions rather than questions. Here are eight pretty good things to say about your new product idea:

1. There's a real need for this product.

2. The improvement over what exists is obvious, apparent, and beneficial.

3. The consumer will be sold on its advantages without special education.

4. The market potential is large and lucrative.

5. The market is easily identifiable.

6. The market can easily be reached.

7. The tooling costs are in line with the profit potential.

8. The spread between the true cost and the optimum selling price will allow for an acceptable profit.

Naturally, you'll add or subtract points to fit your particular product, but by the time you're finished, you'll hope to have Stu nodding his head in agreement so fast that he'll get a headache.

But wait—if you have the time, go to the business section of your nearby bookstore and look at all the books available on salesmanship. There are

Everybody Likes a Good Story

Once upon a Time, I Had a Terrific Idea

tons of them, and each is written by a super salesperson who guarantees to turn you into one. Every book promises to reveal the one secret of sales success that has never before been uttered in public, but the truth is that all these books tell you the same secret. For instance, you'll find the secret buried on page 127 of Herbert "Flash" Gordon's book, *Sales Stories from Hell.* It's on page 91 of Norman "Genghis" Kahn's book, *Selling Door to Door in Beverly Hills.* And you'll find it revealed on page 142 of Wilbert "Whiz" Banger's book, *Famous District Sales Managers throughout the Ages.* Flash is a great salesperson and so is Genghis. And Whiz is probably more successful than Flash and Genghis put together, so I know that what they have to tell us is true:

The most important secret to sales stardom is to think from the customer's point of view and tailor your presentation accordingly.

Who can argue with this? As one businessperson pointed out, "People don't want quarter-inch drills; they want quarter-inch holes." If you understand what benefits the prospect is *really* looking for in your product or service show how these will be achieved by buying your offering, then you, too, can write a book on salesmanship.

It's a given that Stu is looking for profits. Also, because his company is a half step behind Ajax Outdoor, Inc., it doesn't take a genius to assume that he'd like to catch up to Ajax. With that in mind, and heeding the super salespeople's advice, you can slightly rephrase the features of your new product. Here are now eight very good things to say about your new product idea:

1. Yes, Stu, because there's a real need for this product, you'll be able to get quick, profitable, national distribution.

2. Because the improvement over what exists is obvious, apparent, and beneficial, the sell-through in the stores should be sensational.

3. Because the consumer will be sold on the product's advantage without special education, you won't have to spend money needlessly on advertising.

4. Because the market potential is large and lucrative, you're going to make a fortune, and the folks over at Ajax are going to have a fit.

5. Because the market is easily identifiable, your marketing people won't have to bother with expensive research projects.

6. Because the customers can so easily be reached, you'll be turning over profits before Ajax knows what hit it.

7. Because the tooling is in line with the profit potential, you'll have your investment back in months instead of years.

8. So you see, Stu, because the profit is there, you've got a megahit on your hands, and the Ajax guys will be so angry they won't be able to see straight.

This is called "benefit selling." The other thing any good salesperson will tell you is to *always* ask for the order. If the company representatives don't like your product idea, they won't be embarrassed to tell you clearly and promptly. If they do like it, however, you sometimes have to glean that from signals. Here are a few examples:

You Know It's Time to Close the Deal ...

- When the conversation moves beyond the merits of the invention or product idea itself and onto more practical matters, such as production cost, packaging, and marketing.

- When the person you're meeting with calls another person into the meeting and that person shows enthusiasm for the idea.

- When you see that the prospective licensee has a pleased expression and keeps playing with your prototype with a proprietary air.

- When the other side nods and seems pleased with your answers to questions. This is particularly true if the questions have to do with projected sales volume and profits.

Even when the signals are there for everyone to see, many people simply *hate* to ask for an order. It seems, somehow, to be the most vulgar thing you can do, even though it's understood that's why you're there in the first place. Stu may love your idea but he's finding it hard to say yes ("What with the economy being the way it is, and it's a bad time of year, and money is tight, and like, you know."). What good old Stu needs is a gentle shove. This is when you say, "Stu, to be perfectly honest, I had already made appointments to show this to two other companies. If you tell me we have a deal, and if we can shake hands on it, I'll call them and cancel." Stu will never ask you which other two companies, and he hates the thought that a competitor would make a big thing out of something that he let slip by. It puts some tension in the air and adds a nice sense of urgency to the meeting.

Stu is ready to do the deal, but he wants just another moment. To give him a little more thinking time, you might have to go through this little drama:

STU: So what about knockoffs? How do I know I won't be knocked off before I ship my first order? *(Assuming the product isn't patented.)*

YOU: Stu, if a pro like you ever worried about knockoffs, you'd never produce your first product. By the time the competition sees the item and tools up, you'll have a six months' head start. With your marketing know-how, by then you'll have our product in every decent store in the country, and they'll have to settle for the leftovers.

STU   *(already knows this, but doesn't mind a little stroking):* Yeah, okay, so what's the deal on this? *(The rest, as they say, is history.)*

A CUSTOMER'S TWO CENTS' WORTH

So far, everything we've discussed has been from the seller's point of view. I think it's worthwhile to let a buyer express an opinion. Tory Bers is a top product developer with one of the leading toy companies. Part of her job is to interview people with ideas to present, so I asked her for a few pointers. Here are some of the things she mentioned:

1. *Make Sure Your Presentation Doesn't Look Shopworn.* If your presentation is wrinkled and dog-eared, it's obvious that it's been shopped around quite a bit. Nobody is going to say yes to a proposal that so many people have already rejected. The presentation must always look crisp and new, and every prospect must feel that he or she is the first person to see it. Surprisingly, I don't recall ever being asked by a prospect if this was the premier presentation. I imagine they assume it is because I always work with fresh-looking material.

2. *Don't Oversell.* Make your presentation once. Do it thoroughly and with enthusiasm, but don't get too aggressive. Be truthful and straightforward. You're dealing with experienced, professional businesspeople, so if you push too hard, you're going to make ene-

mies instead of friends. When in doubt, pause to listen. Many times that will get you further along than talking.

3. *Be Flexible, Easygoing, and Helpful.* If you come across as rigid, unpleasant, and bullheaded, the prospect might fear that there will be problems in dealing with you down the road. If a customer is on the fence about your product, your attitude may be the deciding factor, so show your agreeable side at all times. The first lesson any salesperson learns is how important it is to be nice. When the caveman Zoggo, the world's first professional salesperson, first started in his career, he used to hit his prospects over the head with a rock to get their attention. Zoggo sold prelit bonfires, and sales in his territory were terrible. Finally, one day, instead of his usual beaning of the prospect, he smiled and asked about the kids. The other caveman had never actually seen a smile before, but he knew he liked it, and it generated warm feelings toward Zoggo. Sales for his prelit fires soared, and soon Zoggo was able to move his family into a charming three-chamber cave, where they all lived happily ever after. And since that time, over the years, the lore has been passed on from salesperson to salesperson: "You'll get more with a smile than with a whack on the head."

4. *Come to the Meeting Appropriately Dressed.* I read in the paper about a scientist who recently made a major discovery concerning the big bang theory of the creation of the universe. When he made the announcement to his colleagues, he did it while wearing a tuxedo to underscore the importance of the event. A man after my own heart!

   Businesspeople are like anyone else; they like to deal with people who look like them and seem to share their values. If the meeting is in a big downtown office and you walk in wearing nice clothes and well-polished shoes, you flatter your host by indicating, "This meeting is important to me." Underscoring the importance of the event, by inference, adds stature to the new product you're presenting. Someone who looks important surely must be selling an idea that's important.

   Conversely, if you're meeting someone in a factory who comes to work in jeans, you'll make that person uncomfortable if you show up in a three-piece pinstripe suit. I had such a meeting last year and deliberately wore a sport shirt and slacks. The individual I met with actually thanked me for not wearing a tie.

How you dress is a dramatic form of nonverbal communication. You tell people in an instant what you think of them and what you think of yourself. If they like what they see, it makes the rest of your job easier. If they don't, it's going to be uphill for you all the way. Although it may appear superficial, what you wear is noticed, and it matters.

This is not the place to make a fashion statement. Earrings are out for men; low-cut blouses and tight skirts are out for women. Chewing gum is always a mistake, and I don't even want to discuss smoking. People look at you and form an immediate and enduring impression. I've read that in a social situation, these impressions are made in about four minutes ... but in a business situation they can be made in as little time as *15 seconds!* But don't worry, with a little attention to your grooming and a happy smile on your face, the first impression you make is bound to be a good one.

If you were applying for a job and a secretary told the director of human resources that you were waiting in the lobby, the director might want to know what you look like. What the director really wants to know is whether you look as if you'll fit in. Like it or not, that's what it's about if you are a salesperson. Fitting in is half the battle.

5. *Cultivate Good Speech Habits.* I know I don't have to caution you about cursing. You needn't spend money on a book to tell you that. However, other patterns of speech that you may not have thought about are equally offensive to some ears.

All these personal points—how you look, how you dress, how you speak—may seem trivial in relation to selling a new product concept, but I can assure you they're not. At either end, none of this matters. If your new product idea is fantastic, its merit will come through even if you make your presentation while wearing a gorilla suit. If your idea is really bad, you can wear a tuxedo and talk like a member of the House of Lords and still not make a sale. Most new product ideas, however, fall somewhere in between and require a degree of skill and effort to get them placed. And the first prerequisite of good selling technique is to create a favorable impression with your customer. The prospect who likes and respects you is more inclined to do business with you. That's where good grooming and acceptable speech come into play.

For instance, excessive use of slang or teenage idioms sounds silly coming from an adult. It trivializes you and your product idea.

Also, you run the risk that your customer isn't going to know what you're talking about. Always speak standard English and stay in step with your audience. If you are young and your customer is old enough to be your parent, you don't want to talk about the "modern" way of doing things and to thereby imply that your customer (the old dinosaur) may not understand. If your customer is thinking, "That little pipsqueak ... " while you're making your presentation, then you've obviously lost the first round.

On the other hand, if your customer is much younger than you are, don't point up the age difference by constantly talking about how things used to be done before her time. Your customer will resent the implication, and you will only have made things more difficult for yourself. The idea is to defuse the age difference and, to the extent that you can, make this a meeting between two contemporaries who have come together to discuss an exciting new product idea. We don't talk politics, sex, religion, or race. We're here to take care of business. If you must make some small talk, discuss sports or the weather.

When you realize the value of first impressions and how quickly they're formed, you can turn it into a wonderful advantage. If you really concentrate on the first few minutes of your meeting, you can have the other side thinking you're bright, witty, charming, cultured, informed, intellectual, and all in all a very classy act. By the time they start to form a few second impressions, you might be out the door with a contract in your hand.

WHEN ALL IS SAID AND DONE

You have now completed your brilliant presentation, and you handled all of these questions with flair and intelligence. You flattered everyone in sight, and they all loved your outfit. You made them laugh, and you made them cry. So now what happens? Now they tell you one of four things.

1. *"Thanks, but this product really isn't for us."* You naturally should accept this decision with good grace, but press for further clarification. Are they really saying it's not a good idea, period? If so, can they suggest ways to make it better? If it's truly just not for them, can they suggest someone it is for? Can they give you a name? an introduction? There's benefit to be gained even from refusals, but you may have to dig for it. I long ago learned not to take these

rejections to heart, and they don't upset me. It really may not be a good product (it won't be my first), and I either improve it or put it aside. Mickey Rooney once said that you always pass failure on the road to success, and I know that's true for me and for just about everyone else who's out there trying. A professional attitude is to understand that there is nothing personal in these rejections. If you keep this in mind and believe in your idea, none of it will get you down. Your idea truly may not be right for this prospect, but it may be perfect for the next one.

2. *"I like your idea, but I have to show it to my boss."* If it's at all possible, you want to avoid leaving your material in the hands of a screener. If a presentation is going to be made to the person who makes the decisions, you should be the one to do it. So say something like, "I'll tell you the truth: I'm so excited about this product idea that I'd like the fun of showing it to your boss myself. But I wouldn't want to do it unless you were there, so I hope you can set something up for the three of us. I realize how busy you are, so any time that's best for you is okay with me." Flattery often works wonders. In fact, you could make a case for the statement that flattery works in almost any kind of situation. People never tire of hearing nice things about themselves, no matter how outrageous they are. With a little practice, you'll soon be able to do it with a straight face. So do whatever it takes, but don't let the assistant make the presentation for you. I don't recall *ever* being successful when I didn't make my own presentation to the decision maker.

3. *"Can you leave the stuff for a while so we can study it?"* You have one of two situations developing here, and it's important to distinguish between them.

   A. If they want to keep your material for a week or so to discuss it among themselves, this is perfectly understandable. However, it's perfectly in order for you to establish a firm date by which time they agree to give you a response. It's important that date be clearly agreed to because you want to maintain a sense of urgency and you don't want them to dawdle. If they do, chances increase that they'll lose interest in the project. When they ask for time, say, "I certainly can understand that you want to study it. I would, too, if I were you. There's a lot to think about. However, as you can imagine, I'm excited about the idea and am anxious to get it placed. So can you tell me when I can expect to have an answer?" Their request is in order, as is your response.

B. If they want to keep your material for several months so they can build a sample and price it and test it and show it to their salespeople and to their key customers, that's another story. What they want is an option, and options are bought, not given. After all, you're giving up quite a bit by cooling your heels for all this time while they put your product through its paces. You certainly are entitled to be paid for your troubles. It's called a *holding fee.* The sum is negotiable, but the concept is that the money is never returnable. However, if they decide to proceed, it's reasonable that it should apply against advances or future royalties.

However, the deal is arranged, it's not a good idea to leave your material on the spot. You should always tell them that you'll send it. There are three reasons:

A. You want to leave a paper trail, so your cover letter confirms the meeting and lists the material being turned over for study.

B. The letter gives you the opportunity to once again stress the points made at the meeting and to reconfirm the date by which time they have promised to give you an answer.

C. Try earnestly to avoid turning over original material. Color artwork can inexpensively be reproduced, so it is almost as good as the original. Even the most well-meaning people may lose your work, misplace it, or throw it out. I've had expensive prototypes accidentally tossed in the trash on two separate occasions. Both times I billed the companies, and both times they paid with profuse apologies. But you never get back the article's actual cost, and you end up losing time.

As I mentioned previously, I always submit my material by FedEx or by the Post Office's Express Mail service. It's in keeping with the sense of urgency and importance that you want to maintain over the whole procedure.

4. *"Okay, you've got a deal."* If you're dealing with a small company, you'll probably be working directly with the owner, who is an entrepreneur accustomed to making quick decisions without committees. In that situation, it's not unusual for the owner to be ready to do a deal on the spot. So if you come prepared with a blank contract in your briefcase, you may leave with a check in your pocket. It happens more often than you might think.

In regard to making multiple submissions, there isn't a single correct answer. I can only tell you my system: I find the company that I think is the very best prospect for my new product idea, and I present it to that firm exclusively. Many times, it's the number two company in the business, on the theory that number one is fat and contented, whereas number two may be a bit leaner and more receptive to new ideas. I tell them honestly that I've shown the product to no one else and that they're the people with whom I want to do business.

If they turn me down, I make multiple submissions to other companies with abandon. I sometimes have a proposal for the same product under consideration by three different companies. If company A buys it, I tell companies B and C that I am withdrawing the proposal. Interestingly, the next time I offer products to B and C, I usually get very quick responses.

After the hard work you did to create and develop your product idea, it's a thrilling moment when a company executive offers a hand and says, "Let's do the deal." It may not happen on your very first call, but if your idea is sound and you persist, it's going to happen sooner or later. It's harvest time, and the right contract will earn you all the fruits to which you're entitled.

6

REAPING THE HARVEST

Realizing Maximum Profits from Your New Product Idea

Whoever said money can't buy happiness didn't know where to shop.

—Anonymous

HIGH ROADS AND HIGHER ROADS

You can profit from your million-dollar idea in four basic ways. Some are riskier than others, and some hold out a higher profit potential than others.

1. You can form a company to produce and market the product yourself.

2. You can sell your idea outright to another company.

3. You can license your idea to another company.

4. You can, perhaps, license the use of your idea to several companies.

Market It Yourself

I don't want to attempt to dissuade you from pursuing this route because we all know of garage operations that went on to make the owner rich and famous. Bill Gates, for instance, comes to mind. However, it's important to know what you're getting into.

First of all, it's important to know that the potential marketability of your product idea is far down the list of important ingredients for success—and if that's all you have going for you, failure is almost certain.

If you ask a Harvard marketing professor to name the five most important ingredients for a successful business, he'll list them in this order:

1. Adequate capital.

2. The knowledge and ability to run a business.

3. Boundless energy and an entrepreneurial spirit.

4. An affinity, love, or instinct for the industry or the product category

5. A marketable idea

The principle is that regardless of how good the product idea might be, it's not going to matter if you don't have enough capital to sustain the business until it starts making a profit or you don't have the skill to run a business properly or you're not willing to put in the hours to make it work or you don't have a natural instinct for the industry. On the other hand, even if the product idea is mediocre or plain lousy, having sufficient financing along with the other skills and knowledge mentioned gives you the means and time to convert a bad idea into a good one.

Without all the ingredients in place, statistics say it's likely that your business will fail by the third year. Assuming you can raise the capital, it'll probably mean pledging your house and other personal assets. Some have the stomach for doing that, and others don't. I've done it myself when it worked, and I made a ton of money. I've also done it when it didn't. I'm a born entrepreneur, accustomed to taking risks, but when you lose your house, it's a crisis that you never forget.

We all know of new companies that have grown and prospered; it still happens in the United States, and that's what makes our country great. Nevertheless, the overwhelming percentage of patented new products are never marketed, most new products that are marketed fail, and most companies that are formed to sell them go bankrupt. It's commonly estimated that only 1 in 100 start-up ventures has adequate financing, and within a few years, most of them are gone. When I get a million-dollar idea, I'm happy to let an established, successful company market it. My house is safe, my assets are safe, and I'm not ducking any creditors. My share of the profits is smaller, but I prefer to take the money and run.

Sell Your Idea Outright

This is a judgment call, and you'll only know after the fact whether you were right or wrong. Obviously, this is the safest of all the choices, and whether it's right for you depends largely on how entrepreneurial you are. It also depends on how urgently you need the money, how much confidence you have in the idea, and your level of faith in the company you've appointed to market it. Because we're already aware of the failure rate of new products, it's not necessarily foolhardy to accept a decent cash settlement and move on. But you never know.

The man who developed the concept for the G.I. Joe doll sold it outright to Hasbro, the big toy company, for $100,000. He recognized that boys liked to play with dolls just as girls did, so his idea was to develop a line geared to their tastes. At the time, more than forty years ago, this was a revolutionary concept. His total investment was a few dollars to buy some samples to show Hasbro what he had in mind, the idea was considered quite risky, and $100,000 was serious money back then, so it must have seemed like an incredible coup to be paid so much for so little. How was he to know that Hasbro would proceed to sell hundreds of millions of G.I. Joe dolls, plus many millions more in G.I. Joe ships and planes and tanks and guns and you name it? How could he possibly anticipate that he would have earned royalties on billions of dollars in sales? And how could he have possibly guessed that more than forty years later, sales would still be going strong?

In another case, way back in 1898, a man named Joshua Lionel Cowen invented an electric flowerpot. It had a battery and a little bulb built in, and when you pressed a button, the flowers lit up. Not surprisingly, there wasn't a very big market for light-up flowers back then, so he sold the idea outright for a few dollars to his dear friend, Conrad Hubert. Hubert, being nobody's fool, immediately threw away the flowers and the pot they came in. He kept the battery and the little bulb and made a new contraption that he called an "electric hand torch." He started a new company to manufacture this product and named the company Ever-Ready. Do you think it's reasonable to assume that his friend Joshua might have regretted selling his idea outright for a few dollars? If he had licensed it instead, his heirs might still be collecting royalties.

To offset these stories, there are probably many more tales of companies that bought product ideas outright for large sums and were never able to make a go of them. Because the products failed, the stories are harder to track down, but you can be sure it happens every day. All you can do is weigh the facts available at the time, trust your instincts, and pray that you make the right decision. No one can expect you to do more than that.

License Your Idea to an Established Company

This is the most usual route and is the one I prefer to take. Over the years, I've had offers from companies to buy my ideas outright, and I've always refused. Sometimes I was right, other times I was wrong, but on balance I'm positive I've earned more money by licensing than I ever would have by simply selling the ideas. My contract always includes an advance payment so I have some compensation if the idea fails (which is not uncommon). And if it's successful, I can hope to enjoy some of the fruits of my creativity. We'll look at contract arrangements in detail later in this chapter.

Multiple Licensing

Although very few ideas qualify for licensing in this manner, it's potentially the most lucrative method of all. You license your idea to one company and still have it to license to another and another after that. For as long as it lasts, it's a perpetual motion money machine. Obviously, your idea would require strong legal protection, either patents, copyrights, or trademarks, and must have industry-wide significance. If companies believed they couldn't compete successfully without access to your idea, they would line up to get a license. If you've developed something like this, congratulations. You've hit the jackpot!

Introducing One of My Foremost Failures

The most common area for multiple licensing is what is known as "character licensing." Spiderman is a character. Mickey Mouse is a character. A few years ago I developed a concept for character licensing that I named "Toga Tales." It consisted of cartoon depictions of Roman-type busts on pedestals, complete with garlands and togas, except they were all hippos in assorted colors.

Each one had a make-believe Latin name and a recitation of his deeds. For instance, Billus Maximus was the first Roman attorney to win a IV-figure settlement in a chariot accident case, and Flunkus Maximus was the first Roman teacher ever to give a pop quiz. And I had Auditus Maximus for accountants, Flossus Maximus for dentists, and several other characters. The idea was that these were to be awards for the "World's Best Teacher," the "World's Best Lawyer," and so on.

It was not the most brilliant licensing concept ever developed, perhaps because it was too complicated, but I did manage to sign up several manufacturers. The products they made with my license failed miserably in the retail stores, but because of the advances, I still managed to make a little money. That's the incredible power of character licensing. It's even possible to make some money if the idea is terrible. Character licensing has made fortunes for the owners of hot properties such as Superman, Batman, Barney, and the Simpsons, to name but a few, so I was determined to try again. I now have a new idea, a new concept, which I'll soon be ready to show. And if this doesn't work either, I'll try again. You have to simply shrug off the failures, knowing that if one idea clicks, it'll more than make up for them.

Every June an event called The Licensing Show is held at the Jacob Javitz Convention Center in New York. People with ideas like mine show their concepts to manufacturers who may want to put these designs on their products. Thousands of manufacturers attend, running the gamut from beach towel companies and lunch box producers to children's vitamin manufacturers, and each one is willing to pay great sums for a character license that captures their fancy.

As you might guess, this is not an easy field to break into, but the financial rewards can be enormous. If you'd like more information, resource names and addresses are included in the Appendix.

THE DOTTED LINE WALTZ

Because this book is about licensing, I'm going to assume that you will not opt to produce your product yourself and that you're not going to sell it outright. I'll also assume that because your product idea is so clever and your presentation was so brilliant, the other side's ready to do a deal. You're not going to fall apart now. If you follow the next few steps, you'll breeze through the negotiations and leave the office with a smile on your face and a big check in your pocket.

The Gentle Art of Negotiating

A reporter for *The New York Times,* for a Sunday article, visited 12 stores on upper Broadway on the west side of Manhattan. None were chains and none were supermarkets. They were all individually owned. In 10 of these stores, she was able to negotiate a better price than that shown on the sales ticket.

"I love these shoes but they're more than I planned to spend. If you could give me 10 percent off, I'd take them right now."

"This dress is perfect, but it needs that scarf to go with it. I can't afford to pay for both—so if you could throw the scarf in for free, I'm ready to write a check."

I made up this dialogue, but I'm confident it resembles what transpired. It's negotiating. If you'll do this, I'll do that. It doesn't have to mean shouting, yelling, or grinding someone into the dust. In its best form, it's simply coming to a meeting of the minds. Okay, so maybe you hope it'll be a little more beneficial for you than for the other person. Nevertheless, it's a time-honored craft, and certain productive techniques have been developed over the years. Here are seven of the most important ones:

1. *Know What You Want before You Start.* Before you even show up for your appointment, you should have a clear idea of what you want out of the deal. Each of your requests should be reasonable, well thought out, and defensible. By reading this chapter, you'll know each of the areas where the licensee may make demands, and you'll know how to address them. Every time you hesitate or seem uncertain over one of the contract points, the other side will be tempted to take advantage. It's like the old saying about the other guy promising to meet you halfway, all the while thinking that he's already standing on the dividing line. It's not difficult to prevent this from happening. If you know all the steps, the dance can be fun!

2. *Don't Be Greedy.* I asked a half-dozen patent attorneys to name the biggest stumbling block in successful licensing negotiations, and they all said it was unreasonable demands from the inventor. To be fair, I think if I asked the same questions of a half-dozen inventors who had potential licensing deals that didn't materialize, I'm sure they'd say it was due to the posturing or ineptitude of their attorneys. Regardless of who's at fault, although there are certain points on which you can't compromise, it's vitally important that a sincere

effort be made to understand the other side's point of view. If you can appreciate the customer's true concerns, you can often accommodate them without sacrificing your own. In negotiating, as well as in selling, those who have mastered the art of listening have a real edge. It's pointless for you or your attorney to continually make arbitrary demands that the other side can't concede. If you keep asking for the moon, you'll eventually get the gate.

An old Italian proverb says, "It's better to lose the saddle than the horse." The sample contract later in this chapter offers a clear, sensible balance between the two sides. It can guide you through the process to a satisfactory conclusion. And you'll still have the horse you rode in on.

3. *Don't Get Personal.* You must develop a proper perspective about the negotiating process. Whatever transpires, it's not life or death; it's just business. Whatever the outcome, you'll still have your health, you won't go to jail, and no one is holding your firstborn for ransom. Therefore, it's never excusable to raise your voice or to be impolite. Amateurs may do that, but professionals never yell and never curse. They're never brusque and never rude. If you push, the other side will shove, and you'll soon have a brawl on your hands. What good can come of that? The purpose of negotiation is to resolve conflict, not to create it. People so obviously respond better to kindness that you have to wonder about the true motives of mean-spirited negotiators.

4. *Don't Be a Patsy.* You must be prepared to make concessions because it's part of the process. However, you can turn this into an advantage if, on principle, you don't concede anything without a struggle. If you simply agree to whatever's asked of you, it unfortunately emboldens the other side to ask for more. You can halt this escalation by meeting each demand with one of your own. "Okay, if I agree to Point B, will you see it my way on Clause 6?" Fair's fair. "Okay, I'll do this if you'll do that" is the essential spirit of negotiating. It's not about one side dictating surrender terms to the other.

5. *The Fear Thing.* John F. Kennedy once observed that we should never fear to negotiate, but we should also never negotiate out of fear. If you're prepared to walk if your minimum demands aren't met and if the other side understands this, then you have a powerful bargaining chip. As you'll see from the following sample agreement, your minimum demands are so reasonable that you shouldn't want to do business with anyone who won't give them to you. It's a

portent of bad things to come, and you'll be better off taking your invention or new idea elsewhere. You must never forget that people who can create commercially profitable new products are in demand, and if this company wants your product, there are others who will as well. When you accept this as truth, it shows. You'll feel your power, and so will the other side.

6. *Win by Listening.* The other side is at the table because the customer's convinced that your idea will make money. She *wants* to do business with you; she *wants* to make the deal. However, just as you have certain core demands, so does she. There's no reason these demands have to be in conflict. If a customer says (or implies) "Take it or leave it," don't pay any attention and don't dig in your heels.

   Your job is to listen carefully to get a clear understanding of what the other side really wants and why. Once you have this insight, it's often possible to restate the licensee's demands in a manner that you can accommodate. Or you can offer "additional information" that will allow a face-saving shift in the customer's position. You're not out to beat them; you're out to accommodate them. The better you listen, the easier it is.

7. *Don't Forget about Tomorrow.* Now that you're on the verge of licensing your new product idea to this company, there's a good chance that you'll be back with another new idea. If you create the impression that you're bullheaded or unreasonable, the company may think twice about wanting to go through this process with you again. Stick up for your demands but always be pleasant.

Just as some of us, in our roles as salespeople, are embarrassed to ask for the order, many of us are also embarrassed or fearful to negotiate on our own. To some, it seems pushy and unseemly; to others, there's the fear of looking foolish or not doing it well. These are attitudes based on a misconception of what the negotiating process is all about. It needn't be a debate or a trial, and there aren't supposed to be winners and losers. If you understand clearly what reasonable demands you must obtain and why you deserve them, you won't be a loser. And if you accede reasonable demands to the other side, your customer won't be a loser either. Paraphrasing Hannibal, you either find a way or you make one.

Many people, including myself, enjoy the process very much. I never allow it to get personal and look at it as an opportunity for a frank, spirited, intellectual discussion. By the time you have finished this book, you'll also be able to do it well, and perhaps you will also enjoy the process.

DO YOU USE AN ATTORNEY OR NOT?

The correct answer to this question is "I don't know." You have to do what your budget allows and what makes you comfortable. Naturally, if you are negotiating a major license involving great sums of money, it's prudent to have professional assistance. However, legal help costs thousands of dollars, and in most instances I honestly believe you won't get a contract as good as or better than the one presented here. My contract is a deal closer. When the prospective licensee indicates a desire to go forward, my contract will allow you to leave with a done deal, signed and delivered. Although I wrote this agreement myself (and I'm not a lawyer), it's been vetted by dozens of lawyers over the years and tweaked a thousand times, but the core points have never been successfully challenged.

Clearly, it would be bad advice to push you into negotiating without an attorney if you feel uncomfortable doing it on your own. After you've

**To the Licensee, You're Starting to Look Like Attila the Hun,
and He's Suddenly Bearing a Striking Resemblance to Der Führer**

completed this chapter, however, you'll see that the contract requirements aren't all that complicated. If it's a simple deal and the other side doesn't have an attorney involved, you may feel confident that you can handle it on your own.

If negotiations reach a point where you and the licensee each have attorneys sitting by your sides, the nature of your relationship starts to head south. Your attorney begins with "What if . . . " and then proceeds to introduce a clause to prevent some dastardly deed that's liable to be perpetrated by the other side. His attorney counters with a recitation of the heinous act that you're sure to do as soon as the licensee's back is turned. And back and forth they go. You can't blame the attorneys; that's what you hired them to do. But the licensee sitting across from you, a guy whom you used to like, starts to look like Attila the Hun. And in his eyes, you're beginning to bear an uncanny resemblance to Der Führer. The chance that this will ever become a signed deal is now hovering around 50 percent, while the attorney's fee clocks are ticking merrily away.

WHO PROVIDES THE CONTRACT?

When our fictitious Mr. Stewart Bunyon of the Universal Camping Equipment Company says, "Okay, so what's the deal on this?" it would be terrible if you replied, "I don't know. I'll have my attorney draw up a contract and we'll get back to you." The only thing I can think of that would be worse is if you replied, "I don't know. Why don't you have *your* attorney draw up an agreement?" I shudder at the thought. Here are two reasons:

1. If you wait for the other side's attorney to take care of the agreement, the deal's liable to grow cold long before you get it. Terrible things happen fast and usually without warning. The person you're dealing with gets fired, a competitor introduces a similar product, the company's marketing plans change. You have to assume that whatever can go wrong will go wrong. And it's all going to happen while you wait for the licensee's attorney to finally get around to drawing up an agreement.

2. If the guy's attorney draws up the papers, I guarantee you it will be the Agreement from Hell. You'll be lucky if you're not demanded to give the licensee an advance as a gift just for marketing your new product.

When Stu asks you what the deal is, you have to be prepared to tell him on the spot and, if he's ready, to do the deal on the spot. You should go to the meeting with a contract in your briefcase, just as salespeople carry order forms. And just as salespeople have to be prepared to tell customers the cost of whatever they are selling, you have to be prepared to tell your customers the same thing. They're entitled to know, on the spot, what kind of deal you're looking for.

If Stu asked me what I want, as I'm putting my agreement on his desk, I'd say: "It's a simple deal, Stu. I get $_____ advance against _____ percent royalty. The deal's in effect for as long as you sell the item. There's no minimum guarantee, and you can cancel at any time." In just a few sentences I've stated the heart of the agreement and told Stu exactly what he wants to know. How much up front? What's the royalty rate? What's the obligation? How long's the deal? How does he get out of it? That's it. Everything else is detail.

Just a Simple Agreement

What follows is the agreement I use for my own licensing deals. I made it up myself. It wasn't prepared by an attorney. I have never lost a deal because of it, and I have never lost a court case in suing for my contractual rights. It's battle tested and will enable you to do most straightforward deals on your own. If you should decide that you'd prefer using an attorney, that's okay. Understanding the contract will make you a better client.

Date: _____

LICENSING AGREEMENT

Harvey Reese Associates, located at _____ (hereinafter referred to as LICENSOR) has given _____ located at _____ (hereinafter referred to as LICENSEE) the exclusive production and marketing rights to his new product concept as herein described and as per drawings, patent applications, and/or prototype samples previously submitted. In exchange, LICENSEE agrees to pay LICENSOR a royalty in the amount and under the terms outlined in this Agreement.

PRODUCT DESCRIPTION:

1. ROYALTY PAYMENTS. A _____% (_____ percent) royalty, based on net selling price, will be paid by LICENSEE to LICENSOR on all sales of subject product line and all subsequent variations thereof by LICENSEE, its subsidiaries, and/or associate companies.

    The term "net selling price" shall mean the price LICENSEE receives from its customers, less any discounts for volume, promotion, defects, or freight.

    Royalty payments are to be made monthly by the 30th day of the month following shipment to LICENSEE'S customers, and LICENSOR shall have the right to examine LICENSEE's books and records as they pertain thereto. Further, LICENSEE agrees to reimburse LICENSOR for any legal costs he may incur in collecting overdue royalty payments.

2. TERRITORY. LICENSEE shall have the right to market this product(s) throughout the United States, its possessions, and territories, Canada and Mexico. It may do so through any legal distribution channels it desires and in any manner it sees fit without prior approval from LICENSOR. However, LICENSEE agrees that it will not knowingly sell to parties who intend to resell the product(s) outside of the licensed territory.

3. ADVANCE PAYMENT. Upon execution of this Agreement, LICENSEE will make a nonrefundable payment to LICENSOR of $_____ which shall be construed as an advance against future earned royalties.

4. COPYRIGHT, PATENT, AND TRADEMARK NOTICES. LICENSEE agrees that on the product, its packaging and collateral material there will be printed notices of any patents issued or pending and applicable trademark and/or copyright notices showing the LICENSOR as the owner of said patents, trademarks or copyrights under exclusive license to LICENSEE.

In the event there has been no previous registration or patent application for the licensed product(s), LICENSEE may, at LICENSEE's discretion and expense, make such application or registration in the name of the LICENSOR. However, LICENSEE agrees that at termination or expiration of this Agreement, LICENSEE will be deemed to have assigned, transferred and conveyed to LICENSOR all trade rights, equities, goodwill, titles or other rights in and to licensed product which may have been attained by the LICENSEE. Any such transfer shall be without consideration other than as specified in this Agreement.

4. TERMS AND WARRANTS. This Agreement shall be considered to be in force for so long as LICENSEE continues to sell the original product line or subsequent extensions and/or variations thereof. However, it is herein acknowledged that LICENSEE has made no warrants to LICENSOR in regard to minimum sales and/or royalty payment guarantees. Further, LICENSOR agrees that, for the life of this Agreement, he will not create and/or provide directly competitive products to another manufacturer or distributor without giving the right of first refusal to LICENSEE.

5. PRODUCT DESIGNS. LICENSOR agrees to furnish conceptual product designs, if requested, for the initial product line and all subsequent variations and extensions at no charge to LICENSEE. In addition, if requested, LICENSOR will assist in the design of packaging, point-of-purchase material, displays, etc. at no charge to LICENSEE.

However, costs for finished art, photography, typography, mechanical preparation, etc. will be borne by LICENSEE.

6. QUALITY OF MERCHANDISE. LICENSEE agrees that Licensed product(s) will be produced and distributed in accordance with federal, state and local laws. LICENSEE further agrees to submit a sample of said product(s), its cartons, containers, and packing material to LICENSOR for approval (which approval shall not be reasonably withheld). Any item not specifically disapproved at the end of fifteen (15) working days after submission shall be deemed to be approved. The product(s) may not thereafter be materially changed with approval of the LICENSOR.

7. DEFAULT, BANKRUPTCY, VIOLATION, ETC.

  A. In the event LICENSEE does not commence to manufacture, distribute and sell product(s) within _____months after the execution of this Agreement, LICENSOR, in addition to all other remedies available to him, shall have the option of canceling this Agreement. Should this event occur, to be activated by registered letter, LICENSEE agrees not to continue with the product's development and is obligated to return all prototype samples and drawings to LICENSOR.

  B. In the event LICENSEE files a petition in bankruptcy, or if the LICENSEE becomes insolvent, or makes an assignment for the benefit of creditors, the license granted hereunder shall terminate automatically without the requirement of a written notice. No further sales of licensed product(s) may be made by LICENSEE, its receivers, agents, administrators or assigns without the express written approval of the LICENSOR.

  C. If LICENSEE shall violate any other obligations under the terms of this Agreement, and upon receiving written notice of such violation by LICENSOR, LICENSEE shall have thirty (30) days to remedy such violation. If this has not been done, LICENSOR shall have the option of canceling the Agreement upon ten (10) days written notice. If this event occurs, all sales activity must cease and any royalties owing are immediately due.

8. LICENSEE'S RIGHT TO TERMINATE. Notwithstanding anything contained in this Agreement, LICENSEE shall have the absolute right to cancel this Agreement at any time by notifying LICENSOR of his decision in writing to discontinue the sale of the product(s) covered by this Agreement. This cancellation shall be without recourse from LICENSOR other than for the collection of any royalty payment that may be due him.

9. INDEMNIFICATION. LICENSEE agrees to obtain, at its own expense, product liability insurance for at least $2,000,000 combined single unit for LICENSEE and LICENSOR against claims, suits, loss or damage arising out of any alleged defect in the licensed product(s). As proof of such insurance, LICENSEE will submit to LICENSOR a fully paid certificate of insurance naming LICENSOR as an insured party. This submission is to be made before any licensed product is distributed or sold.

10. NO PARTNERSHIP, ETC. This Agreement shall be binding upon the successors and assigns of the parties hereto. Nothing contained in this Agreement shall be construed to place the parties in the relationship of legal representatives, partners, or joint venturers. Neither LICENSOR nor LI-

CENSEE shall have the power to bind or obligate in any manner whatso-ever, other than as per this Agreement.

11. GOVERNING LAW. This Agreement shall be construed in accordance with the laws of the state of _____ (your home state). IN WITNESS WHEREOF, the parties hereto have signed this Agreement as of the day and year written below.

_____   _____
LICENSEE   LICENSOR
DATE:_____   DATE:_____

This is as brief as it's possible to make an agreement and still cover all the important points. Everything is spelled out in clear, layperson's language, with the provisions and obligations reasoned and logical. It's a businessperson's agreement, designed to be understood and negotiated by you and the licensee in a friendly, professional manner. I once showed this to a patent attorney friend who told me that he could easily turn it into a tightly worded, 10-page document, but it wouldn't necessarily be any more effective than it is right now. Let's look at it more closely, section by section.

Introduction

In this section I identify myself and the licensee by name and address. I confirm my right and claim to the new product concept and I describe it for the record. I keep this description as brief and general as possible. It should be specific enough to absolutely identify the product but broad enough to encompass all the derivative products that may spring from the core idea.

Royalty Payments

This is the heart of the contract and there is no correct figure. It's usually in the 5 percent to 10 percent range, with high-volume items earning less royalty than slower-moving products. Items such as toys, appliances, and textiles will usually be 5 percent or less. Products such as giftware and home furnishings are closer to 10 percent or more. If you can glean some understanding of the licensee's profit structure, it'll help you to arrive at a figure. Whatever you think is right, add a point or two to it. You may have a pleasant surprise.

Smaller licensees usually don't object to making monthly royalty payments, much as they pay salespeople's commissions. Larger companies, however, and those that license many products are usually set up to make quarterly payments. This is not unreasonable, and I'm prepared to concede the point. Anything less frequent than quarterly payments, however, would be unreasonable. "Stu, the royalty is my paycheck. You don't get paid twice a year, so why should I?"

Some licensees will say they won't pay royalties until they get paid by their customer. I explain that I'm not in the credit business, and that's one of the reasons the licensee gets 95 percent of the proceeds and I get only 5 percent. In line with the principle of not making a concession without asking for one, I might suggest that I'll accept this payment method if he's willing to increase the royalty to 6 percent. That may only be one point, but it's a 20 percent increase.

It's important that the term *net selling price* be clearly understood and agreed to. You don't need hassles later on when the company sends you your first check. Simply stated, whatever the net amount of the money received by the licensee from purchasers of my product is what I expect to be paid on. The licensee can make any number of deals, but your royalty should be a percentage of the customers' invoices. Don't accept any other definition.

Some licensees ask to have salespeople's commissions also deducted, but I would never agree to that. If you let that go, what's to prevent the company from asking to deduct salespeople's travel expenses or trade shows or catalogs? Remember that your royalties should always be tied strictly to sales. Let the licensee worry about profits—that's why he gets the big bucks. Sales figures are absolute, whereas other figures, such as profits and expenses, can be fairy tales.

My definition of net selling price says nothing about bad debts. Licensees almost never bring it up, but if they did, they would have a point. After all, if their customer goes bankrupt, shouldn't they be allowed to deduct any royalties you perhaps have already received? That sounds logical, but even if the customer goes bankrupt, the vendor often receives something in a settlement. And after your royalty has been deducted, the licensee will never pay you anything back from a settlement. Although I deliberately leave this out of the agreement, I'm willing to concede the point if it comes up. Fortunately, it almost never does, but make the concession if you must. It's too small a point to be a deal breaker.

This provision also allows me to be reimbursed for legal fees if I have to sue the licensee to get paid. Sometimes a licensee's attorney will stipulate that his client is obligated to pay only if my suit is successful. I concede this point if it arises. I understand the intent is simply to head off frivolous suits. It's not a big deal, so you can be generous. This is really a terrific clause for our side. I've successfully exercised it a few times over the years when licensees started to forget to send me my checks. The licensee can't object to the clause's inclusion because that would be an admission of intent to skip payments. In fact, I'd be wary about dealing with any potential licensee that did raise objections.

This clause was a later addition to the contract after I had a collection problem and found my legal fees were almost as large as the sum I finally collected. There are people out there who will simply refuse to pay because they think you won't go to the expense of bringing suit. However, when they're obligated to paying your legal bill, they usually have a miraculous change of attitude. Dorothy Parker once remarked that the two most beautiful words in the English language are *check enclosed*. This clause can help a lot to make that happen.

Territory

Unless the licensee has a strong export program, I usually limit the license to the United States, Canada, and Mexico (now considered a contiguous market). If you have foreign patents on your product and it's already being successfully distributed in the United States, foreign companies will find you to request a license. If you don't want to wait for that, you can contact the trade departments at the embassies of the countries you're interested in, and they'll be happy to provide you with the names of likely candidates whom you can contact directly. However, don't attempt a licensing deal with an offshore company without having legal representation in that country. This is one instance where you definitely can't do it yourself.

On the other hand, if your licensee is an international company, why not let them go all the way? What you can do is establish a separate export sales goal, and if the export department doesn't reach it, you can have the right to remove that portion of the territory. Licensees will agree because they won't want to be bothered with foreign sales either, unless the numbers are significant. If you do award foreign rights to the licensee, refer back to the Foreign Patents section. If you don't have foreign patents yourself, this is the time to request that the company file on your behalf.

Advance Payment

As with the royalty percentage, the amount of the advance payment is highly negotiable. Because there are no rules to follow, you can make up one of your own. "Stu," you ask innocently, "how many utensil sets do you think you can sell in a year?" Stu will probably give you an inflated figure, assuming it will make you more willing to make concessions. If he says, for instance, that he'll sell a million dollars' worth and assuming the royalty has been set at 6 percent, you might say, "That means I can look forward to earning $60,000 the first year. Because advances are often set at 25 percent of anticipated income, suppose we follow the formula and set ours at $15,000. Is that okay?" (I made up this 25 percent rule, but it's as good a way as any to arrive at a figure.)

Stu may counter by proposing a $15,000 guarantee, with half of it to be paid as the advance when the contract is executed. It's your decision. Either way you'll wind up with a minimum of $15,000 and hopefully much more if the product is successful. Whatever the figure, there must always be an advance, and it must always be nonrefundable; if you don't have that, you don't have a deal.

Every once in a while, I encounter a company that simply *hates* to pay an advance and insists that it's company policy not to do it. I'm adamant on this issue. I explain that this agreement is nothing more than taking an option to produce a product. The licensee pays an advance to earn the right to tie up your product idea for months and months. If the company doesn't ultimately proceed with it, the advance is the penalty. The licensee wouldn't be able to get an option on a piece of property without money, so why should it be possible to do it with your new product concept?

Sometimes prospective licensees, desperate to avoid the advance, will *swear* that they are definitely going ahead with the project, so there's no need to pay an advance. If that happens, remember the words of H.L. Mencken: "It's hard to believe that a man is telling the truth when you know you would lie if you were in his place." The bottom line is that you shouldn't sign a licensing agreement without an advance. No matter what the company policy may be or what the licensee swears, if you can't get it, my advice is to pack your things and leave. Almost everything else is negotiable, except an advance.

Amazingly, in my own experience, at least 25 percent of the contracts I sign do not result in the product actually being produced and marketed. The developmental lead time is so long (anywhere from six months to a year or more) that anything can happen along the way. Perhaps the company's marketing goals change or the company gets sold or your idea becomes obsolete or the concept proves to be too difficult and expensive to produce. Whatever the reason, six months later you wind up with the product back in your hands to sell to someone else. I've sold some new product ideas three times to different companies, and the products have *still* never been produced! At least the advance provides some consolation.

Copyright, Patent, and Trademark Notices

This is straightforward and will meet with no resistance. You're simply requiring licensees to put the appropriate legal notices on the product. This is for their protection as well as yours, so they'll do it routinely.

If the idea you presented is in the patent pending stage (or not even applied for), you can negotiate to have the licensee pay the costs to acquire the patent in your name. The same thing applies to appropriate trademark or copyright protection. The argument is that the licensee's stake in the product is much bigger than yours, so it's ultimately to the company's advantage to get all the protection possible. The firm may have house counsel or attorneys on retainer who can handle the legal work in a routine manner. My agreement doesn't force the issue but provides the option. The larger companies usually do it; the smaller companies usually don't.

Terms and Warrants

If your product is not protected by a patent or an unusually strong copyright, it becomes public domain as soon as it's distributed. At this point, it's usually not possible to take the product away from the original licensee to reassign it elsewhere. If another company wanted to, it could simply produce its own version (a knockoff) without seeking your permission or paying you a royalty. Therefore, in this case, licensing for a specific time period is moot. Because my products usually fit into this category, my agreement makes no provision for time period exclusivity; however, that doesn't mean that yours can't.

If the product you're licensing has sufficient legal protection, it certainly makes sense for you to place a time limit on the license. The licensee is entitled to an original license of at least development time plus 12 to 18 months. It's only fair to give the firm sufficient time to recoup the initial investment. Beyond that, if the company is not performing up to pre-established goals, you should have the right to look elsewhere. If goals are being met, the original agreement should automatically renew for another 12 to 18 months, and new goals should be established for review at the next renewal period. Notwithstanding all of this, the licensee retains the right to cancel whenever it appears unprofitable to continue.

Assuming you have not established a time frame, my agreement is designed to remain in effect indefinitely. If the licensee continues to produce your product for 100 years, the royalty checks should still continue. Every month, when the mail carrier brings a royalty check, your rich great-grandchildren will talk about what a terrific genius you must have been.

The main reason this can happen is that the agreement encompasses your original concept plus all new products that may stem from it. If your original product idea has sufficient merit to remain viable over the years, it will undoubtedly be subject to continual change and improvement. Eventually, you may not be able to detect its resemblance to the original, but it's still your product, and you're entitled to continue receiving royalties.

Product Designs

This clause is specific to my situation and therefore may not be applicable to yours. I have certain art, design, and marketing skills that I'm prepared to lend to the licensee to advance the salability of my product. The clause describes what I'll do free and what I expect to get paid for. You may have other unique skills or knowledge that will be beneficial to the licensee and ultimately to yourself. For instance, it may be necessary for you to spend time on the factory floor, providing some technical training to the licensee's production people. Will you do it free? Do you expect to be paid? This is the clause where you make these matters clear.

Quality of the Merchandise

The intent of this clause is obvious; you don't want to see any junk on the market with your name on it. It's harmful to your reputation and will probably lower your royalties. On the other hand, you have to understand that the finished product will *never* look as nice as you pictured it in your dreams. That's why the clause says that your approval cannot be reasonably denied. If the licensee is already in production and you don't provide your approval because you feel the mauve should be a little deeper, the company will sue you back into the Stone Age. It's the same as ordering wine in a restaurant. You can't return it just because you may not like the taste. You can return it only if it's pure vinegar.

Default, Bankruptcy, Violation, and So On

A licensee who is losing money on your product has the right to say, "Enough's enough!" and throw in the towel with no other obligation than to pay you whatever royalties are due. This has happened to me more times than I like to remember, but there are never any hard feelings on the licensee's part or mine. We both understand that new product introduction is a gamble, and as astute as we like to think we are, we guess wrong quite often. Licensees must have the right to back out whenever they want, or they'll never sign the deal.

Indemnification

Picture this scenario. You've created this terrific new product idea and licensed it to a Fortune 500 company. Sales are fantastic and the money is rolling in. A Bentley in your garage is starting to look like a prudent investment. One day, a middle-aged housewife in Snowshoe, North Dakota, claims that the product burst into flames, burned her house down, and inflicted third-degree burns over 40 percent of her body. Three personal injury attorneys were killed in the stampede to her bedside. The surviving lawyer is suing everybody who has ever had anything to do with this product, including you.

"But hey!" you cry, "I just dreamed up the product. I didn't manufacture it!"

"Tough darts," replies the lawyer, "I'll see you in court."

You tear apart your desk, looking for the agreement. "Where is it?" Your wife says the last time she saw the file it was in the basement, under the supply of cat litter. You race downstairs and throw cat litter bags in every direction. Your two cats stare coldly, promising a blood oath to get revenge. *Where is it?* Finally, the folder! As you frantically flip through the pages of your agreement, your eyes finally rest on Clause 10: Indemnification. A wave of relief passes through your body as you see the certificate naming you on a $2 million insurance policy. It's only then, while wiping your sweaty hands off on your Ralph Lauren leisure suit, that you pause to reflect on what a true genius you are.

No Partnership

This simply makes clear that you have an arms-length agreement with the licensee that's limited strictly to the terms of the agreement. If the management get arrested for smuggling illegal labor in from Guatemala, that has nothing to do with you.

If the licensee is from another state, it's beneficial for the agreement to be governed by the laws in your own state. Legal costs are less if you have to sue, and judges and juries tend to favor the local party.

The Sins of Omission

There's a method to providing the licensee with a deceptively simple agreement. Several important points have been deliberately omitted. Their inclusion is not to my benefit, and the omissions are almost never noticed. However, if the points do come up, I'm prepared to address them.

Infringement Protection. Suppose you have a pretty strong patent for your Whatzit, which you've licensed to Amalgamated Global Things, Ltd. Your product is on the market, just starting to sell well, when Amalgamated's archrival, Ajax Trans-World Bettergoods, Inc., knocks off your product and introduces its own Whatzit. Amalgamated is screaming bloody murder and wants you to sue Ajax for the gold in the president's teeth. What do you do? A lawsuit might cost you $100,000 in legal fees.

You never, ever should sign a contract that obligates you to sue. That way can lead to financial disaster. And yet if you don't sue, Amalgamated can say, "So why should we be paying you a royalty if you're not willing to protect us from skunks like Ajax?" On the other hand, why shouldn't Amalgamated sue? After all, they're getting 95 percent of the proceeds, and you're getting only 5 percent. And besides, you're only a single person and your licensee is a big company. So suppose you both decide to sue. How do you split the costs? And if you win, how do you split the proceeds? I have the perfect answer to all these problems. My solution is to duck the issue altogether and worry about it only if Ajax really does knock off Amalgamated.

I've seen clauses in contracts that do address the issue, but the statements are so wimpy that they might as well duck it, too. The clause usually says that licensor will have the right to sue. If it doesn't, then licensee will have the right to sue. Licensor agrees to help licensee, licensee agrees to help licensor, and they all live happily ever after. As a practical matter, if a knockoff does occur and it's important enough, Amalgamated will sue Ajax on its own, without asking for your participation.

Licensor's Title. Some contracts have a clause like this:

> LICENSOR will indemnify and hold harmless from all damages, costs and expenses, including reasonable attorney's fees which may be paid or incurred by the LICENSEE by reason of any claim arising from any breach of LICENSOR's representation or warranties under this Agreement and the LICENSOR, at his own expense, will defend and protect the LICENSEE's approved use of the product.

Let's assume you honestly believe the idea for the Whatzit came from your own creative brain. The licensee believes you and signs the licensing agreement. The firm tools up and starts to manufacture and distribute the product. Some time later, it receives a notice that it is being sued by Ajax, which claims that its own Whatzit patent is being violated. The suit asks $2 million in damages. Amalgamated turns to you for defense. Do you catch a plane to Singapore or what?

Ajax may just be causing trouble, but who knows? In any event, the suit still must be answered. Amalgamated says, "You got us into this. Now we expect you to get us out." Unfortunately, the company has a legitimate point. The best thing to do is take the coward's way out by omitting the point from the agreement. Voltaire once claimed that he was ruined twice in his lifetime—once when he lost a lawsuit and once when he won one. So duck this question if you can. Personally, I would never sign a contract that obligated me to take any form of legal action. I suggest that you don't either or your nights of restful sleep will be over—it's easy to imagine yourself spending $100,000 in legal fees to protect $25,000 in royalties.

Patent Improvements. As time goes on, you may find ways to improve the features of your product. Normally, you would routinely pass these on to the licensee, particularly because they may improve sales and thereby your royalty. But suppose you hate the licensee and are counting the days until the license expires. Do you want the firm to have the improvements? Morally, I suppose the licensee is entitled to them. But I don't see anything in the agreement about it; do you?

Disposal of Merchandise. Let's assume that after three years, Amalgamated decides to cease selling the Whatzits because sales have dropped off dramatically but the company still has $200,000 in inventory. You can't expect the licensee to just dump this merchandise in the river. If it is sold to a closeout business, below cost, should you still expect to receive royalties? And how much time does the licensee have to get rid of the merchandise? It can't drag on indefinitely because you may want to place the license elsewhere.

I leave this out of the agreement also because there's no benefit for me to include it, and it's just one more possible point of contention. And almost nobody ever thinks of it. Nevertheless, if it does come up, here are some guidelines to get through it quickly:

> It's not unreasonable to give the licensee 90 days to get rid of the stock. Less than 60 days doesn't seem fair, and 180 days or more is unreasonable. Nobody ever argues with 90 days.

> If the licensee is getting rid of the inventory at or near the regular prices, you're obviously entitled to receive your regular royalty. On the other hand, you and the licensee should be able to arrive at a formula to reflect a decrease in royalties proportionate to additional trade discounts. For instance, if the licensee gives an extra 15 percent discount, perhaps your royalty should drop by the same 15 percent. With an extra 30 percent closeout discount, the royalty drops by 30 percent. And so forth.

And what about charge backs on royalties? For instance, suppose you received royalties for a big shipment your licensee made to one of its customers and that customer subsequently went bankrupt? Because the licensee was never paid, shouldn't it have the right to collect the royalties back from you? You might say, "Yeah, but credit's your job. Why should I be penalized if you made a bad judgment call?" Why make this an arguing point when the matter may never come up?

Regardless of all the clauses and negotiations and sallies back and forth, the best contract is the one that's reasonably equitable to both sides and after being signed is put in a drawer and forgotten. If you know clearly what you want, the negotiation doesn't have to be unpleasant. In fact, I enjoy that part of the process. I hope this chapter makes you comfortable enough to enjoy it as well. To guide you on your way, here is a summary of the more important negotiable and nonnegotiable demands.

WHAT'S NEGOTIABLE AND WHAT'S NOT

Aside from the boilerplate provisions, the agreement has 20 essential elements, half of which are nonnegotiable. They should be in your agreement as stated, or else you're giving too much away. The other half require some discussion and negotiation, and you should be clear on your

A Few Mild Requests

position before proceeding. A firm, positive approach almost always heads off problems.

I maintain that the principal reason licensing negotiations often end badly is because the inventor and the attorney, too, if the inventor has one, both lose sight of the true purpose of the negotiation. I've always believed that the end purpose is, of course, to wind up with an agreement that gives me what I'm reasonably entitled to—but to do it in a way that recognizes the legitimate entitlements of the licensee and, most important, to use the negotiations as a method of establishing the kind of amiable, working relationship with the licensee that will encourage him to view my next product idea favorably and to make the next contract signing merely a formality.

I'm not an inventor—I'm in the inventing *business.* I intend for the guy sitting across from me at the negotiating table to be a longtime customer. If I let ego get in the way and start to think my new product idea is sandwiched somewhere between world peace and a cure for cancer, the relationship I'm seeking will never happen. And if I used an attorney and allowed her to battle each point to the death, then it won't happen either. A license negotiation is not a take-no-prisoners war. If both sides don't win, then *nobody* wins.

When I sit down to work out an agreement with a potential licensee, the negotiation seldom takes more than a half hour—an hour tops. As the licensor, I always provide the contract. Because I wrote it and I've used it so many times, I know by heart what 10 points I must be prepared to negotiate, and I know exactly what 10 points are nonnegotiable. What I'd like to review first are the ten nonnegotiable points.

When a union goes into negotiation with a list of nonnegotiable demands, no one takes them seriously. They know it's merely an opening gambit. However, for a licensing agreement, the demands I'm about to lay out are deadly serious. You are absolutely entitled to all of them—and being denied *any* of them is enough reason for you to walk away from the deal. I've done that several times and never regretted it. There is sweet reason to justify why each and every one of these points should be granted to you without argument. If the licensee balks, it's time to reevaluate your relationship.

Ten Nonnegotiable Demands

1. *Royalty percentage should always be based on sales—never profits.* It sounds logical when the licensee offers to share profits with you,

but proceed down that path at your own peril. Profits can be creatively interpreted—one way for the tax man, another for the potential investor, and another for you. Sales, however, are absolute, detectable, and irrefutable. Whatever profit the licensee claims or denies for his own purposes is his business. Your concern is sales, and that's what royalties should always be based on.

2. *Sales are sales. Period.* There is only one acceptable definition of *sales*—and that is whatever the licensee's customers pay her when they purchase your product. I have had licensees demand to deduct costs for things like catalogs, trade shows, commissions, travel, and so on, and I always say, "No way!" That's a slippery slope, and you should never get on it. Sales are sales. Period.

3. *Retain the rights to examine the books.* The profits that a licensee makes on the sale of your new product are his business—but how many he sold and at what price is your business as well. Not only must the agreement specify your right to look at the sales records, but if you have to sue the licensee to collect royalties due, *the licensee must pay your legal fees.* You shouldn't have to pay a $20,000 legal fee to collect $5,000 in royalties. Overwhelmingly, licensees are legitimate and will therefore not object to this clause. If you come upon one who does, I suggest you think twice about entering into a relationship with him.

4. *Define your product properly.* The agreement will naturally call for a definition of the product for which royalties are to be paid, and it's vital that you do it properly. Products evolve and change over the years, and you're entitled to receive royalties notwithstanding the fact that the product five years down the road bears scant resemblance to the product the licensee started with. When defining the product, be sure to add "and all subsequent changes and variations thereof" or else an unscrupulous licensee will easily find loopholes to cut you off from the royalty stream.

5. *No end to royalty payments.* "It's very simple, Ms. Licensee: As long as you're selling my product or variations thereof, I expect to be paid royalties. I don't care if it's 1 year or 20—if you're still selling it, I expect to continue getting paid my small share." I've had licensees want to put a cap on the number of years they have to pay royalties, and if I couldn't change their minds, I ended the negotiation. If someone else is making money from your creativity, why should there be a limit on how many years royalty payments are to be made? Fair's fair.

6. *Product approval.* You must have the right to sign off on the product before it's distributed to the public. Not only is your reputation at stake, but there are also safety issues (lawsuits) that must be addressed. This is a great responsibility, and you must understand that you cannot withhold approval simply for esthetic reasons. However, if the product is out-and-out unsafe, you must have the right to withhold approval until the problem is corrected.

7. *The nonrefundable advance.* I will never sign an agreement unless it calls for a check to be put into my hands immediately thereafter. It's to be nonrefundable and is considered as an advance against future earned royalties. In principle, it's what in some circles is called "earnest money." It should be large enough to represent a significant commitment to your idea on the part of the licensee yet not so large as to become a stumbling block. Without the nonrefundable advance, you're at the licensee's mercy. I strongly believe that you should never sign a licensing contract unless an advance is part of the deal.

8. *You must have a date-certain when your product will be on the market.* My standard agreement states that the licensee must have my product on the market in six months or I have the right to cancel the agreement and take my product elsewhere. The advance is mine to keep for my troubles. Six months is arbitrary, and I'm prepared to lengthen it if the licensee has good reason to need more time. What matters is the principle. A licensee can't simply let the project drag on and on without subjecting himself to consequences. Without the advance and without the performance date, you don't have a contract.

9. *Product liability insurance.* No sane company would sell products to the public without plenty of product liability insurance. The assumption is that if something can go wrong, it will, and if there's even the remotest chance for someone to injure herself by using the product, that someone will find a way. If that happens, the lawyer for the injured party will look to sue everybody connected to the product, including *you.* It's not expensive for you to be added by name to the licensee's insurance policy, and the licensing agreement should legally obligate him to do so.

10. *No obligation to sue.* It's entirely reasonable for a licensee to say, "Suppose a year from now we discover that we're infringing on someone else's patent. Will you protect us?" Or, particularly if your product is patented, he might say, "Suppose one of my lousy competitors tries to knock us off. Will you go after her?" One of my ad-

vantages in supplying my own agreement is that I don't address these issues and the licensee rarely brings them up. But sometimes he does, and I have to tell him point blank that I cannot and will not contractually obligate myself to go to court as a blanket obligation. I tell him that it's a matter to be addressed if and when the situation arises and that circumstances will dictate the action that he and I should take. We're partners, I assure him, and we'll do what's best for both of us.

Does that sound like doubletalk? Well, it is. There's no faster way to personal bankruptcy than to obligate yourself to take legal action to protect your licensee. It's easy to imagine a scenario where you'll spend $200,000 in legal fees to protect $20,000 in royalties. You owe it to your licensee to have made every reasonable effort to make sure your idea is free and clear before bringing it to him. But you should *never* guarantee anything beyond that. If you do, a good night's sleep will be a thing of the past.

Certainly there are other points to be discussed that apply specifically to your own invention or new product concept—but if you enter the negotiation firm in your conviction that you are morally and reasonably entitled to these 10 principles, then you are far along in winding up with an agreement that both you and your licensee can live with for many years to come.

Ten Negotiable Demands

In these areas you can't simply dictate terms. You have to respect the licensee's point of view and be prepared to be as accommodating as you can without sacrificing the core positions you're entitled to.

1. *What percentage of royalty will you get?* As previously discussed, the percentage is entirely open for negotiation. In principle, slower-moving products, like giftware, offer higher margins to the licensee, and you should be entitled to a higher royalty percentage. Faster-moving products, like toys or electronics, with smaller profits built in usually demand that your royalty percentage will be lower. Generally, when products are first introduced, the profits to the manufacturer are higher and start to come down only when competition appears. You might ask for a larger royalty percentage at the time of the product's launch, with the provision that it be adjusted downward if and when the licensee is forced to lower his selling price.

2. *What is the extent of the territory being awarded?* The licensee, of course, is entitled to the United States, Canada, and Mexico. If she asks for the world and if she has a strong international department, I would give it to her. However, there should be a performance guarantee built into the agreement.

3. *What is the payment schedule?* My standard agreement says that I'm to be paid monthly, but it almost never stands. I leave it in deliberately. I know that for bookkeeping purposes, my licensees will almost assuredly insist that payments be made quarterly. I always agree "with reluctance," secretly happy to have little things like this as ways to show my cooperative spirit.

4. *What's the term of the agreement?* It's rarely a subject of disagreement, because most licensees understand that as long as they sell the product, they have to pay royalties. On occasion, however, I have encountered licensees who insist on a limit as to how many years they'd be willing to pay. But why should they? Stick to your guns. You have logic and tradition on your side.

5. *Is there a minimum guarantee on royalty payments? How much?* If the licensed product is not patented, you can't demand a minimum payment because once your product is on the market, it no longer can be licensed to another company. They'll just knock it off on their own if they want to. If your invention is patented, however, then it can be moved, and you have the right to demand a certain level of performance. How much? There is no absolute answer and no magic formula. You get what you negotiate for. Sometime prior to reaching this stage, the licensee will probably have given you an estimate of what he thinks he can sell. Setting the minimum at, say, 60 percent of this estimate seems like a reasonable figure.

6. *Do you supply your expertise free? If not, what's the charge?* I'm a professional product designer, so my agreement clearly spells out what services I'm prepared to offer for free and which ones I expect to be paid for. If what you've invented is a new type of machinery, for instance, and if it will require weeks of your time to teach the licensee's production people how to use it, this is where the two of you determine if that's part of the deal or if you're entitled to be compensated separately.

7. *How big is the nonrefundable advance?* This is totally negotiable with no precedent to refer to. I set mine instinctively based on the size of the company and the value I think there is in my product. However,

I never lose sight of the fact that it should represent only a small fraction of the royalties I expect to receive, and its principal purposes are to establish the seriousness of the licensee and to reward me for wasted time if months down the road, the licensee decides not to go forward with my product. That's not an infrequent event.

8. *How much time does the licensee have to bring the product to market?* My printed contract says six months, but I'm prepared to be reasonable if, because of my product's seasonality or complexity, the licensee needs more time. Because the company is investing funds to develop the product, it's in the company's interest to have it ready in a timely manner.

9. *Who pays legal fees to complete patent work?* If your product was licensed as having provisional patent status or no patent at all, it's not unreasonable to ask the licensee to assume the legal costs to have the product fully protected. Larger firms, with patent lawyers on staff or on retainer, can often be persuaded that it's in their best interest. Smaller firms will probably fight you, saying it's your responsibility as the inventor and licensor—but there's no harm in asking.

10. *Who pays legal fees in case of lawsuits?* The matter does come up from time to time. However, as previously discussed, I urge you not to be swayed. You can't promise to initiate legal action, and you can't promise to defend your licensee against legal action. Your worst nightmares might come true if you do.

Congratulations! Now that you have an executed agreement and a check in your pocket, it's time to go home and start working on your next million-dollar idea.

7

THE WONDERFUL WORLD
OF THE INTERNET

How Did Inventors Get Along without It?

*The future belongs to those
who believe in their dreams.*
—*Eleanor Roosevelt*

As mentioned in the Introduction, when thinking about writing this new edition, I realized I had three reasons why it could offer a worthwhile advancement over the previous one. First were the new options afforded to inventors by the Patent Office, the most important being the Provisional Product Application Program. This was discussed in some detail in Chapter 3.

The second reason for this new edition was the fact that over the past few years I've been in contact with more than a thousand private inventors—affording me considerably more insight into their thoughts, fears, needs, and problems than I had before. I've made a special effort to address these matters in appropriate passages throughout the book, and those who read the earlier edition will have no problem detecting new bits of information and a refocusing of emphasis on almost every page.

And finally, the third reason for the new edition is the incredible rise in importance of the Internet over these past half-dozen years. Its impact has been felt by everyone, not the least of whom are my fellow inventors, innovators, and idea people. Suddenly, the work we do is not so lonely any more! It's easy to join a newsgroup with other inventors when you just feel like relaxing or talking shop. Stuck? Got a problem? Have a question? Need a resource? Need an expert's advice? No question is too esoteric for the Internet if you phrase it properly and search hard enough for the answers.

Simply go to any of the major search engines, enter a proper word or phrase about the invention process, and a bewildering number of sites will appear. Visiting many of these sites might be a waste of time, but others might be quite valuable and informative. Here's my personal list of the ones that I believe deserve your attention:

Patents, Trademarks, and Copyrights

www.uspto.gov General Number USPTO

www.uspto.gov/web/office/com/iip/index.html Independent Inventor Program

www.uspto.gov/web/menu/pats.html Patent Information

www.uspto.gov/web/offices/pac/provapp.html Provisional Patent Program

www.uspto.gov/web/offices/pac/disdo/.html Disclosure Document Program

leweb.loc.gov/copyright Library of Congress Copyright Office

Before spending time and money on your product idea, the logical first step is to make sure it's original. Aside from marketplace searches, an obvious stop should be the Patent Office web site. You can conduct a free search for any patents that have ever been issued involving your idea, which are in force, and which have expired. Also, on the Patent Office site you can order booklets, find out about the various programs, check on fees and filing requirments, and download necessary forms. You can even make applications for patents and trademarks online. As well, if you have copyright questions, a visit to the Library of Congress web site will be beneficial. If you're Canadian, you can get similar services at dipo.gc.ca/.

Scientific Assistance

For those of you working on scientific or energy-saving inventions and looking for information or grant money, you might find it profitable to visit one or both of these web sites:

www.oit.doe.gov Department of Energy Inventions and Innovations

www.nsf.gov National Science Foundation

Sources to Locate Potential Licensees

I believe the best place to find potential licensees is in the marketplace itself, but if that doesn't work, you can search the following databases for the world's manufacturing companies. Please, I implore you, select ones that you can visit *personally.*

www.thomasregister.com The Thomas Register of American Manufacturers

www.mfginfo.com Manufacturers Information Network search engine

www.asiansources.com Asian Sources Directory of Asian Companies

Trade Shows

With your invention under your arm, a trade show is a wonderful place to identify potential licensees, make contacts, and show your idea to companies that you could not otherwise visit personally. Any of the following web sites will help you find the shows in which you're interested:

www.tscentral.com

www.cyberexpo.com

www.expoguide.com

Invention Evaluation

Any expert will tell you that you should do two things before visiting lawyers or otherwise spending money on your product idea. First, make sure it's original, and second, have it reviewed by an outside professional to determine if it has sufficient commercial potential to warrant your time and financial investment. Any company that offers a "free" evaluation should be avoided. The people there will positively tell you that your idea is brilliant so that they can proceed to sell you their other services. The following web sites, including my own, are for legitimate sources that charge a fee, have no other services to sell, and have no vested interest in telling you anything but the truth.

www.wini2.com WIN Innovation Center

www.innovationcentre.ca Canadian Innovation Centre

www.edu/business/innovate.htm Wisconsin Innovation Services

www.sbdc.wsu.edu/innovate.htm Washington Innovation Assessment center

www.Money4ideas.com Harvey Reese Associates

Licensing Agents in the Toy and Game Field

This is one of the few industries in which agents are a major force in the introduction of new products. The following companies listed are all legitimate, with excellent connections in that industry. They all work in a similar manner: You pay them a fee to evaluate your toy idea, and if they like it, they'll proceed as your agent to look for a licensee. If successful, they receive a royalty share. I include my own web site in this group because I also handle toys—but I'm not exclusive to the toy field as the others are.

www.Exceld.com Excel Development Group

www.Maradesign.com Mara Design Associates

www.NewFuntiers.com New Funtiers, Inc.

www.Anjar.com Anjar Company

www.Ayers-concepts.com Ayers Concepts

www.Creativegroupmarketing.com Creative Group Marketing

www.Money4ideas.com Harvey Reese Associates

Agents in Other Fields

As explained elsewhere, it's not easy to find full-time professional licensing agents in most fields; however, here are some who have carved out specialty niches for themselves:

www.Adlenterprises.com/default/html

This is a division of the Arthur D. Little consulting organization and only deals with scientific and high-technology intellectual property transfers.

www.Accessorybrainstorms.com

Accessory Brainstorms works only with fashion accessories and beauty products.

www.Healthcareinventions.com

This company works with medical doctors who have invented medical-related products.

Prototyping Services

I can think of no circumstance where having a prototype at a product idea presentation wouldn't be better than not having one. As the inventor, it's your responsibility to prove that what you've invented will work in the manner you claim it will. Usually, the most effective way of accomplishing this is with a working prototype. If it's not something you can do yourself, there are probably lots of local shops who can do it for you. Here are some professional organizations that offer prototyping services on a national basis:

www.Creativegroupmarketing.com Creative Group Marketing (toys only)

www.Maradesign.com Mara Design (toys only)

www.Protosew.com Protosew (custom sewing)

www.members.aol.com/T2design T3 Design Company (general)

www.Inventionmakers.com Invention Makers (general)

www.Adesigner.com A Designer (general)

Miscellaneous Valuable Web Sites

For their own individual and unique reasons, I believe you'll find visits to each of the following web sites to be informative, interesting, and rewarding.

United Inventors Association, www.uiausa.org. This is a nonprofit membership organization for the education, assistance, and protection of individual inventors. The site is filled with useful articles, programs, and information and offers a venue to ask questions directly to experienced specialists.

The Andy Gibbs Web Site, www.PatentCafe.com. Created by an experienced, successful inventor, The Patent Café would be on anyone's list of inventors' top 10 must-visit sites. The pages are well designed and easy to navigate, and no web site on the Internet has more useful information, services, programs, usable forms, and materials to offer inventors than this one.

Inventor's Digest Magazine, www.inventorsdigest.com. Inventor's Digest is America's only magazine devoted to the needs and interests of private inventors. The web site offers an archive of past articles, information about the current issue, trade show information, a directory of inventor organizations, and inventions for sale along with a variety of other useful information.

The Jack Lander Web Site, www.Inventorhelp.com. Jack, an inventor himself, specializes in the sale of books for inventors. He's read what he sells and can offer reading suggestions in response to your specific needs. Obviously you can buy books online from Amazon.com or Barnes&Noble.com, but you won't have a professional guiding your purchase.

The Inventors Network, www.InventNet.com. Run by Vic Lavrov, this is a general inventor's web site offering memberships and featuring pertinent articles and programs. The site hosts the popular InventNet Forum for on-going discussions about matters of interest to private inventors. If you have an engineering, venturing, or patenting question, this is a good place to ask it.

The National Inventor Fraud Center, www.inventorfraud.com. Unfortunately, lots of companies prey on innocent inventors, and this is the site that points fingers and names names. Operated by attorney Michael S. Neustel, The Inventor Fraud Center sorts through the bad and the good firms. Look over these lists before writing a check to anyone.

The Ronald J. Riley Web Site, www.InventorEd.com. Nobody on the Internet is more passionate about educating and protecting the rights of his fellow inventors. Ron's site, which he rightly calls Inventors Ed, is so full of articles, instructions, and exposures of corrupt invention submission companies that it's fairly bursting at the seams. Ron recently added his personal e-mail forum where some of the wisest people in the field meet to discuss inventing issues. You can actively join in or simply lurk; it's fun either way.

Inventors Newsgroup, alt.inventors. This is an unsupervised free-for-all meeting place for inventors, patent lawyers, patent agents, engineers, experienced inventors, and newbies. The regulars are always happy to welcome new voices and there's hardly a question about inventing you could ask that someone can't answer. The only problem is that, as with many unsupervised newsgroups, it gets its share of unsavory visitors. However, it's still a terrific source for information—either by participating or simply lurking.

This list is far from complete, but the web sites mentioned are intended for general inventor interest and are among the ones that I visit often. However, the Internet offers countless other valuable and interesting sites that may be more focused on your particular needs. If you need a patent attorney recommendation, the name of a local inventor's organization, a company to help with your product design, someone to build your prototype, sourcing information for esoteric materials, agents to help license your idea, companies to conduct expert patent searches, pamphlets on inventing, forms for nondisclosure agreements, sample contracts, free product evaluation forms ... they're all there, awaiting your visit. And looking is free!

THE CURSE OF THE INVENTOR'S WEB SITE

Before closing this chapter, I want to say a few words about inventor web sites. Hardly a day goes by without my receiving at least one e-mail from an inventor urging me to visit her web site and report back on what I think of this fantastic, wonderful new invention. I write back each time to explain that, sorry, I don't visit inventor sites to give snap, prejudgments of new product ideas. I need to have the invention information in front of me to quietly contemplate.

As I say, the e-mail request to visit these sites is a daily occurrence, and it finally started me thinking about the value, if any, of inventor web sites in general. In the back of my mind I've known for some time that I was vaguely against them, and I finally gave some thought to the reasons why.

Bluntly speaking, I honestly believe that a site does the inventor far more harm than good, although I *do* understand the following powerful arguments to support an opposing view:

Argument Number One. It's so easy to do. Almost anyone can make his own web site—and if not, the inventor can always ask the first 14-year-old kid who walks by.

Argument Number Two. It doesn't cost a cent to post it on the Internet. You can have your own homepage, or you can put your page onto any one of lots of existing commercial web sites that invite inventors to post their inventions for free, assuming, I suppose, that collectively the site will draw traffic and perhaps the one who runs it can sell some advertising.

Argument Number Three. Considering that it's free, what's there to lose? So what if it's a long shot? Who knows? Maybe just the company you've been looking for will come upon the site and follow up with a hot phone call. Where's the harm?

Before I tell you where's the harm, you tell me—where's the good? Can you give me *one* instance? I'm in contact with thousands of inventors— and I *never* heard of a case where an inventor made a deal for himself through his web site. I don't claim it's actually impossible or even that it has never, ever happened—I'm only saying I've never heard that it ever did. More to the point, however, in terms of actual harm—I see two really serious problems.

Problem Number One. An inventor creates a web site, proudly sits back, and thinks she's done something wonderful. Full of pride and eager ex- pectation, willing to put her faith in the electronic miracle of the Internet, the inventor finally has an excuse other than simple sloth to not pursue personal attempts to get her idea licensed. After all, posting a site is a lot easier than calling a company president to try for a personal appointment. It's more fun to get pats on the back from friends and relatives whom you've persuaded to look at the site than it is to make a face-to-face pre- sentation to a bunch of grim-faced marketing people at a company two states away.

"No! No!" you may say—"I use my web site as a marketing tool. I don't just ask friends and relatives to look at it—I write to companies and ask them to look, too! It's like showing the idea in person." But who's kidding whom? *Nothing* is like showing the idea in person. And besides, we all know it's really just an excuse to avoid going out into the world to sell your idea.

Problem Number Two. If you're a man proposing to a woman, you're not going to say that you've been turned down three times before and you hope and pray this time you'll hear a yes. Nobody's dopey enough to pro- pose like that ... but you're doing the same thing by trying to market your invention through a web site. You're shopping it around, hoping to finally hear a yes. What you want to say to the woman is that of all the women in the world to whom you could have proposed, she's the special one whom you want to marry—and that's the same message you want to give a potential licensee: Of all the companies in the world that could be interested in this new product idea, you selected the Jones Company because it's such a perfect fit with their existing product line.

Instead, by advertising your invention on the Internet, you're acknowledging that it's already been seen by dozens or hundreds of others, none of whom had the slightest interest—so why would the Jones Company respond any differently?

When I make a presentation to a company, I do everything in my power, short of out-and-out lying, to create the impression that they're the first folks to ever see the idea. As far as they're concerned, I ripped the sketches off my drawing board that morning and the paint isn't dry on the prototype. I *never* go to the meeting with shop-worn presentation material, even if it has to be done over because I've used it twice before. No company wants to know that two or three competitors have already turned the idea down. If your invention's posted on the Internet—you're saying *the whole world* has turned it down. From my perspective, you could hardly do more harm for yourself than that.

By trying to promote your invention on the Internet, you're saying it's an old, stale idea, and as my grandmother once taught me—no fishmonger yells that the fish he's selling isn't fresh.

8

THE UNLIMITED POTENTIAL OF PRODUCT LICENSING

You may be disappointed if you fail,
but you'll be doomed if you don't try.
                                        —Beverly Sills

This is a tough time for American manufacturers. That's what they always say. Foreign competition is fierce, the economy is on shaky ground, and in every area of commercial activity, too many companies are trying to get a piece of the pie. This may be bad news for them, but it presents a wonderful opportunity for us.

BUSINESS STINKS—BUSINESS IS BOOMING

When companies think they're in trouble, what do they do to increase profits? It doesn't matter if it's a mom-and-pop operation or a Fortune 500 company, they all try to do the same thing: Cut costs and increase sales. And what does every worthwhile new product do? It either works to increase efficiency (cut costs) or helps to increase sales. That's no coincidence. It's why licensing is such a booming business and why it gets bigger with each passing year. It's a vital part of the world economy. Without having access to licensed products, inventions, and technologies, many companies would simply cease to exist. If you've developed a terrific new product concept that will cut costs or increase sales, any company executive will welcome you with open arms and a ready checkbook. Just call on the phone and you'll be invited to come right over.

Throughout this book, I've tried not to automatically refer to new product ideas as *inventions* (although they may very well be), and I've tried not to refer to us as *inventors* (although that description might be correct). The licensing business is so much more than just inventors selling inventions that I didn't want to confine your thinking to just that one activity. The following examples make this distinction clear.

Million-Dollar Babies

Years ago, an artist named Xavier Roberts from Cleveland, Georgia, developed a concept for a doll and licensed it to one of the large toy companies. The idea was that each doll would be presented to the child as being newborn, with its own name and birth certificate. The child doesn't simply buy a doll; she *adopts* one. It's a simple idea that any one of us could have had if we had focused our thoughts on that area. You would think that at least one of the doll companies would have thought of it. After all, it's their business, and some of them have been around for a century or more.

With all due respect to Xavier Roberts, it doesn't take a genius to think up ideas like this. It's probably not even an original idea, but Xavier deserves to get all the rewards because he's the one who did something about it. He developed the idea, designed the doll, gave it a name (Cabbage Patch Kids), and went out and found the right company to license it.

Xavier is not an inventor, and Cabbage Patch Kids is not an invention; it's a merchandising scheme. Nevertheless, Cabbage Patch Kids have accounted for more than $3 billion in retail sales, turning an artist from a little Georgia town (who probably couldn't care less about being called an inventor) into an extremely wealthy man.

Cold, Cold Cash

Iowa winters, as you can imagine, are brutally cold, and if you're a farmer starting a tractor at 5:00 A.M., it must make you dream of gentle sands and swaying palms. One local farmer, at least, doesn't have to dream about it anymore. He probably has a mansion in Hawaii.

A few years ago, I had dinner with the sales manager for a Midwest manufacturing company who told me a charming little story. One day, out of the blue, a farmer appeared in the company's waiting room and asked to see the president. Because it was a small company, the farmer was quickly obliged, and he proceeded to demonstrate a product idea that would dramatically change his life and enormously improve the business fortunes of this little firm.

What he took out of a paper bag to show the company president was a store-bought mitten with a little hole cut out of the front. In the hole was shoved a plastic windshield scraper from an automotive store. That's all there was to it. The farmer explained that it was the best way he found to keep his hand warm every morning when he had to scrape the ice off his farm equipment, and he thought other farmers would also find it useful. He reckoned that if the company produced it in an attractive way, farmers would pay good money for it. Obviously, the fleece-lined mittens with the built-in scraper that the company ultimately produced had an appeal that went far beyond Iowa farmers. I can't begin to guess how many millions have been sold, but the sales manager told me that they ship the product to every cold-weather country on the face of the earth. I live in a cold-weather part of the United States, and hardly a Christmas goes by when I don't get one or two as presents. At this point, the product is a staple.

The fact that nobody bothered to get a patent for a mitten with an ice scraper stuck in it didn't deter either the farmer or the manufacturer. They signed a licensing deal, and both are quite happy they met. The farmer didn't have to read this book to understand that anything capable of producing profit is capable of being licensed.

Every time you hear a song on the radio, someone is collecting royalties. Every time you buy a book, someone is collecting royalties. Every T-shirt a kid buys with a rock star's picture on it, or every baseball card, earns a royalty for someone. If you can patent your idea, fine. If you can't, don't worry about it. If it's good, if it's a moneymaker, and if you find the right buyer, it's not going to matter.

How High Is Up?

No one can begin to estimate how big the licensing business is. Certainly, it's in the billions of dollars, but how many billions is anyone's guess. The only fact everybody does agree on is that it's getting bigger all the time. The nice part is that much of this money is going to average citizens with no more talent, brains, or ingenuity than the rest of us. It's going to farmers, waiters, artists, and accountants—all ordinary people with ordinary jobs.

Thousands of new products are introduced every year, while thousands of others are quietly discontinued. Some companies that guessed right have grown to be bigger than ever, and some companies that guessed wrong have slipped into oblivion. However, those of us who earn our living by creating new products are always on the job to serve the newest batch of entrepreneurs as well as the established larger clients. We've risked nothing, so we're still here, quietly collecting our royalties.

There's No Such Thing as Golden Oldies

In our trendy, fast-moving society, new products are being devoured at an ever-increasing pace. Most of the products on the retail shelves today didn't even exist five years ago. New technologies destroy and create whole product categories in a single gulp, while our changing lifestyle makes products obsolete in a heartbeat. Every old product must be replaced by a new one. Companies believe that if they don't grow, they die. So the frantic search for exciting new products is never ending, and it creates the best of all seller's markets for people with appropriate products to license.

The Old Products Graveyard

Among two of the busiest trade shows of the year are the National Close-out Shows in Chicago and the Variety Merchandise Show in New York. These venues are where America's obsolete products come to die, and the entire retail world attends the wake. Walking the aisles, snatching up bargains, you might find a department store buyer from Zaire walking past a discount store owner from the Philippines, who just bumped into a bargain barn operator from Kentucky. Once U.S. manufacturers have decided to discontinue a product line, it's urgent for them to dispose of the inventory as quickly as possible. They need whatever funds they can get out of the product, and they need the space in their warehouse for new merchandise. There's an entire industry of closeout specialists who bid against one another with cash offers for the inventories of distressed manufacturers. You might be astonished how big this business is. And it's all to make room for the new product you've just developed.

I've been involved in several businesses over the years and have been exposed to many more. In my experience, I've never seen anything to compare with licensing for profit opportunity and personal growth advantages, with so little financial risk.

WHY THERE'S NO BUSINESS
LIKE THE LICENSING BUSINESS!

1. *There Is No Competition.* There's *always* room for a great new idea. No one will ever turn you away because they already have enough ideas, and no one will ever tell you that your competitor has a better price. There is no competitor, and there's no such thing as a better price. At the very beginning of this book, I commented that I had no reservations about telling you everything I know about this business. I'm not worried about creating competition for myself. All of us together could never satisfy the constant, insatiable demand for great new product ideas ... it never stops! You'll deal with top management, and every company you license to is a potential repeat customer. You'll always be invited back when you have your next big idea. Boom times or bust times, it makes no difference—the door is always open. Can you think of any other business like this?

2. *Virtually No Investment Is Required.* The literal truth is there is some investment. You may have to pay for a patent, a prototype, or a photograph. But this is pocket change compared with what you'd spend if you attempted to market the product yourself. For the person of average means, the financial outlay required could hardly be described as risk capital. It's hard for me to think of another business that has such a huge profit potential in relation to the capital requirement.

3. *No Experience Is Required.* Over the years, I've created and licensed more products than I can count. I have lots and lots of experience. Nevertheless, the very first product you create could very easily be superior to any that I've ever done. The edge I've gained from experience has nothing to do with the quality of the idea. Who could teach you that? My experience has taught me how to repeat the creative process, and I know tricks to getting my new products licensed. Now that you've read the book, you have all the knowledge you need. The only things you must bring to the party are intellectual curiosity and a will to succeed.

4. *There Is Unlimited Potential.* If you apply for a salesperson job and they tell you it has unlimited potential, what they mean is that if you're good at it and work hard, you can make a good living. When I say "unlimited potential," I'm talking about the sun and the moon and the stars and the planets. Here's what I mean:

HEROES ON A HALF SHELL

In 1983, two Army veterans used their discharge money to start publishing an underground comic book. It was a crude one-color affair that they sold to those little comic book stores you see here and there. It started as a gag, a spoof. To poke fun at the standard comic book superheroes, these two fellows created four turtles named after Renaissance painters. They were dumped into the sewer, as small turtles often are, and years later they emerged as full-blown teenage crusaders, ready to do battle against the forces of vice and evil. They were named the Teenage Mutant Ninja Turtles.

For a long, hard five years, the creators, Peter Laird and Kevin Eastman, continued in this fashion, putting out their slim comic book. Toiling away in their small town of Northampton, Massachusetts, and continually going out to comic book fairs across the country, they eventually managed to build up a small cult following for their "heroes on the half shell."

At about that time, a bright, knowledgeable young man named Mark Freedman left his job with a large licensing company to go out on his own. He named his one-man company Surge Licensing and set out to look for clients to represent. Somehow, he came upon an issue of the Turtles' comic book and immediately recognized the licensing potential in these four little guys. He drove up to Northampton and the rest, as they say, is history. Even in his wildest dreams, however, Mark couldn't have imagined that this would turn out to be one of the greatest licensing bonanzas of that decade.

The three young men—Peter, Kevin, and Mark—joined together in an alliance that has made licensing history. More than $2 billion in retail sales of products with the Teenage Mutant Ninja Turtle logo on them has been recorded, and that's not saying anything about the income from two feature-length movies, record contracts, book deals, foreign licensing, and other projects far too numerous to mention.

I don't want to even try to guess how many millions of dollars each of these young men earned in royalties and fees before the Ninja Turtle fad finally died, but the amount obviously gives new meaning to the words *unlimited potential.*

It doesn't matter if you create a new style of windshield wipers to license to the auto industry or just a poster about *General Hospital.* The right licensee will be delighted to give you thousands or millions for your idea. We see or hear about new success every day. In one issue of one of my trade magazines, for instance, I saw a little article about one of the electronic game companies paying an American inventor $43 million for the right to use his technology in their games. No big deal, just another inventor collecting millions.

In licensing, there's no such thing as reaching a saturation point. It's impossible. The supply of new products and new merchandising ideas can never satisfy the demand. It's a contradiction of terms. The better the new products, the bigger the demand for more. The more exciting the new products, the bigger the demand for more. And on it goes. If this book has awakened you to the possibilities of licensing and made you think about your own potential, I'll be very pleased. Nothing would make me happier than to be partially responsible for creating a new batch of millionaires. That would make this book the finest new product I ever created.

9

MAKING YOUR LICENSING DREAM A BEAUTIFUL REALITY

With money in your pocket,
you are wise, you are handsome,
and you sing well, too.

—Anonymous

I once heard Jimmy Carter explain why he decided to make a run for the presidency of the United States. His answer seemed almost too simple, but it held a powerful message. All he said was that as governor of Georgia, he had met many world leaders, and it had finally dawned on him that he was at least as smart as any of them. Once he realized that, nothing could hold him back.

What holds the rest of us back is our tendency to underestimate our own ability while we're overestimating the ability of others. Once Carter accurately assessed his strengths, his ambition and drive were unstoppable. Surely we can do the same in our own way. The components for business success are reasonable intelligence, some talent, a plan of action, and the self-confidence to see it through. You certainly have the intelligence, and I can assure you that you have more than enough talent. The book provides the plan of action, so if you can just add a measure of self-confidence, you'll be on your way. Just remember what the little engine said while chugging up the mountain: "I think I can, I think I can." You have to see yourself as a success before you can become a success. You can be sure that little engine would still be sitting at the bottom of the climb if all it could say was, "I'm not so sure I can, I'm not so sure I can...."

We are all familiar with the expression that if it quacks like a duck and waddles like a duck, then it must be a duck. But this concept also works for swans. If people act as if they are successful and carry themselves as if they are successful, then they will be perceived as successful. We've all met people who, by the sheer force of their attitude about themselves, can project an image of accomplishment. Their self-confidence comes through and makes them winners.

If you tell me you've developed a commercially profitable idea for a product, I won't be particularly impressed. Thousands do it all the time. If you tell me you've retained a topflight attorney who at this very moment is applying for an ironclad patent, I'll hardly be in awe. The Patent Office is clogged with applications (they've issued more than ten million patents over the years). But if you tell me you have a beautiful prototype, a drop-dead presentation, and appointments to show your idea to three different companies, I'll know I'm talking to a winner! If you have the confidence in your idea and the confidence in yourself to put your drawings under your arm and knock on doors, I know you're going to make something good happen.

> Many have heard opportunity knocking at the door—but by the time they unhooked the chain, pushed back the bolt, turned two locks and shut off the burglar alarm, it was gone.
>
> —Anonymous

All the patent lawyers I've spoken to tell me the same thing. The typical inventor dreams up an idea, applies for a patent, sends out form letters to maybe a dozen companies, and then sits back waiting for lightning to strike. It just about never does, and all the person ends up with is a good-looking patent certificate to hang on the wall and maybe show off to friends. The majority of patents are issued to small, independent inventors, and very few of them are ever commercialized. Inventors often don't seem to have the courage and self-confidence to go out and sell their ideas, and they try to do it by mail because that's easy, safe, and non-threatening. Or worse yet, they give their money to an invention marketing company that promises to do the work for them. The bad news is that these approaches seldom work; the good news is that terrific things happen to people who get out there and do it! It's not that terrible things happen to people who won't make the effort; it's that *nothing* happens. The months and years go by, and every day their dreams fade just a little bit more. And one day, it's just too late. The world truly has passed them by.

WE MAKE OUR OWN LUCK

As you can see, the secret isn't talent or brains. The secret is *action*. When you can get into action mentally, physically, and emotionally and when you focus all that enormous energy into the small hole of a clearly defined goal, you'll take off like a rocket and nothing in the world is going to stop you. Calvin Coolidge once noted that the world is full of smart, talented losers. We all know people like that who are just sitting around in their safe cocoons, complaining about the good luck that happens to other people. They may understand and pay lip service to the idea that people make their own luck, but today there's something they really want to see on television. "And, you know, tomorrow I promised to go fishing with the guys and next week starts the bowling league. But I know you're right, and in the spring I'm really going to get myself in gear." They're planning for failure while the rest of us are planning for success.

It's not difficult to realize why people who may intellectually understand are still unable to shake themselves loose. It's fear of failure. "Flop sweat," they call it in the theater. Many capable, decent people are so frightened of failure that they won't even try for success. A person who polishes and nurtures a dream may fear that by trying to make something of it and failing, the dream will be destroyed. And without the dream, what does the person have? Never trying at all technically keeps the dream alive. The person's deadly lie is to promise to do something about it. There's an old Spanish proverb that says tomorrow will be the busiest day of the year.

FEAR OF FLYING

Fear of failure and lack of confidence are caused by ignorance, and the obvious cure is knowledge. People who are afraid of flying, for instance, are helped by being taught about the workings of airplanes and the principles of flight. The more they understand how an airplane flies and the more they know about all the reassuring backup systems built into modern aircraft, the less frightened they become. And when they know enough, flying starts to become bearable for some and actually enjoyable for others.

My most sincere hope is that this book will work in the same way. By knowing everything there is to know about licensing—how to get an idea, how to protect it, how to get a lawyer, how to get a patent, how to get an appointment, how to make a presentation, how to conduct a meeting, and how to write a solid agreement—you will have the medicine you need to get rid of fear and take action. If everything is known, then there can be no fear of the unknown. You may even find that what you are doing is fun. And there aren't many better things in life than being paid to do work that is fun.

When I set out to write this book, I had three objectives in mind. By providing you with some insight into licensing, I wanted to convince you that

- You should do it.

- You can do it.

- You will do it.

You Really Should Do It!

Demonstrating that you should do it is easy. Success stories about average people getting big payoffs for licensing simple ideas are all around us. A waiter in Seattle developed and licensed a game called Pictionary. So far, more than forty million sets have been sold, and it's still going strong. Think about it. If a waiter can become a millionaire from a simple, clever idea, shouldn't you be wildly intrigued by the possibilities? I can assure you from long experience that manufacturers are *eager* to give you money. Just demonstrate that they can turn a nice profit from your wonderful new product idea. Surely that's not too much to ask. Besides, what have you got to lose? A few dollars? A little time? That's an awfully small investment when you consider the potential rewards. The worst-case scenario is that nobody will buy your idea. It's not the end of the world. You'll have benefited from the experience, and your next idea will be better. The best-case scenario is so good that I'll leave it to your imagination. Either way you're a winner.

You Really Can Do It!

Demonstrating that you can do it is a little more difficult, but I think we've achieved that objective as well. I'm living proof that an average person can win at this game. You win by participating. Achieving your goal is the trophy. Just think how many times an idea of yours eventually appeared on the market by someone else's effort. Suppose it was you who did something about just one of them instead of its always being someone else? Probably you would only have had to do it once.

Yesterday, it possibly could have been said that you have more intelligence and creativity than you know what to do with. But today you know what to do with it. In this book is the plan. You just have to put it to work.

You Really Will Do It! (Won't You?)

I'm not sure if you will do it. That's up to you. Have you ever bought a motivational book? They're usually written by psychologists, and they often sell in the hundreds of thousands. This is not necessarily a testimony to their effectiveness. Rather, it's a testimony to the wide desire among so many people to "finally get started." These books are 300-page pep talks, and on almost every page they say something like *"Do It!"* or *"Go for it!"* But do what? Go for what? They don't tell you that. That's your problem.

It's not the motivation that's lacking; if people were lacking motivation, they wouldn't have bought the books in the first place. What's lacking is a direction, a plan of action. By providing licensing as a blueprint for success, I hope I've provided the flight path. This is certainly not the only way to riches. There are wonderful books about real estate investing, stocks and bonds, franchising, and almost every other commercial endeavor you might care to attempt. Each offers a legitimate opportunity for great success. Licensing, however, is the only one I can think of that involves virtually no financial risk, and the rewards can be just as great. If you don't at least *try*, you'll be denying yourself a unique opportunity. You can have almost anything you want if you work for it and stay focused on your goal.

Licensing is the single best way I know of to earn big-time wealth with small-time risk. People no brighter or more talented than you are earning millions of dollars from their ideas. There is no better time than right now to get started; manufacturers are desperate for new products and new ideas. Nearly every time I sign a contract, the licensee says, "I hope you make a million dollars from this deal." And it's an honest statement, because if I make $1 million, the licensee makes $10 million or $20 million. So nothing would give a manufacturer greater pleasure than to send you a fat check with lots of zeroes each month. The licensee will put many times that amount into its own bank account. So buy yourself a notebook, go back to page one, and get your brain in gear. Somewhere out there companies are waiting to give you thousands and thousands of dollars. All you have to do is bring them products they can make money with. A fellow named Clinton Jones once said, "I never been in no situation where having money made it worse." Let those be your words to live by.

Speaking of words, there's an old tune with words something like: "You can do it if you wanna, but you better know how." This book has given you the know-how. The "wanna" is up to you.

APPENDIX

When I was thinking of the changes I wanted to make in this new edition, one thing I knew for sure was that the book should be oversized to allow the included forms to be usable as well as useful. I hope you like this new format. I've tried to not clutter this Appendix with unnecessary facts and figures—and I've included only material that I believe will help you to achieve great success with your new product idea. It's an eclectic mix, and I think you'll enjoy reading through it.

Also, although I've made every effort to double-check all phone numbers and addresses, I know it's inevitable that some will become obsolete as time passes. In the publishing world, many months go by between the time a manuscript is prepared and the printed book begins to appear in stores—and, of course, more months or even years can pass before the book winds up in any one reader's hands. If the probable obsolescence of any of this material causes you any measure of bother or inconvenience, I offer my apology and ask for your understanding.

CONTENTS

HARVEY REESE ASSOCIATES
Non-Disclosure Agreement

Dear Inventor:

You intend to submit your invention or new product idea to me for evaluation in respect to licensing its use in exchange for royalty payments. If I believe the idea is commercially viable, I will offer my services as your agent in a subsequent agreement, which you are free to accept or reject as you see fit.

In order to allay any fears and to initiate what is hoped will be a fruitful relationship, I offer the following Non-Disclosure Agreement covering this product idea: (identify product)_____ .

Agreement between Harvey Reese (REESE) and INVENTOR known as:

NAME:_____   ADDRESS:_____

INVENTOR confirms that, to the best of his or her knowledge, he or she is the originator of the product idea and has the full power to submit it to REESE for his evaluation.

REESE confirms that he will only show and/or discuss the idea internally, and will not show or discuss it elsewhere without INVENTOR's permission.

All information about INVENTOR's idea shall remain confidential except in the following circumstances:

1. If information about the idea becomes public knowledge through no fault of REESE.

2. If the INVENTOR on his or her own makes disclosure of product information to the public, or if information about the product comes to REESE in good faith from outside sources.

3. If REESE is already working on a similar or identical product concept as demonstrated and proven by REESE's internal records and files.

REESE agrees that he will not use the confidential information for his own advantage without INVENTOR's express written permission.

REESE agrees that at no point does he become a co-inventor, and any design improvements he might make to the product idea will automatically be assigned to the INVENTOR without cost or obligation.

AGREED:

_____   _____
INVENTOR   HARVEY REESE ASSOCIATES

AGENCY AGREEMENT

HARVEY REESE ASSOCIATES
("HRA")
614 South Eighth Street, PB 305
Philadelphia, PA 19147

INVENTOR NAME: _____
("The Inventor")
Address: _____
_____

Invention Name:_____

BACKGROUND

WHEREAS, THE INVENTOR has made an invention idea or design which he or she believes to have commercial potential, but has not yet successfully commercially exploited it on his or her own, and

WHEREAS, HRA also believes the invention has commercial potential, and has a great deal of expertise in the field of marketing new ideas, and

WHEREAS HRA is interested in working with THE INVENTOR to commercially exploit the invention, but only on terms of appropriate compensation to HRA in the event the invention becomes commercially successful;

NOW, THEREFORE, THE PARTIES for and in consideration of the mutual covenants hereinafter provided and other good and valuable consideration, the receipt of which is hereby acknowledged, agree as follows:

I. DEFINITIONS

As used in this Agreement, the following capitalized terms (whether used in the singular, plural or possessive) shall have the following meanings only:

1.1 THE INVENTION means the technical developments and ideas made by THE INVENTOR THAT ARE DESCRIBED IN THE ATTACHED appendix A.

II. OBLIGATIONS AND CONSIDERATIONS GIVEN BY THE PARTIES

2.1 Agent's Duties
HRA shall:

(a) use its best reasonable efforts to promote and extend commercialization of THE INVENTION throughout the United States of America, and, if opportunities shall arise, in other countries throughout the world.

(b) study and keep under review market conditions to ascertain the most likely commercial partners that might have interest in using THE INVENTION or producing products in accordance with or including THE INVENTION.

(c) at HRA's own expense, shall prepare or have prepared on its behalf any prototypes, exhibits, demonstrations or marketing materials that in the reasonable best judgment of HRA may be useful or necessary to advance commercialization of THE INVENTION.

(d) give proper consideration and weight to the interests of THE INVENTOR in all dealings and abide by any reasonable rules or instructions notified in writing by THE INVENTOR to HRA.

(e) not represent any other person whose interests shall interfere with effective marketing and commercialization of THE INVENTION pursuant to this Agreement.

(f) not act in any manner which will expose THE INVENTOR to any liability nor pledge or purport to pledge THE INVENTOR's credit, and

(g) defray all expenses incurred by HRA in the performance of its duties under this Agreement.

2.2 Exclusivity

(a) THE INVENTOR agrees not to appoint any other person to act as its agent for any duties that would likely overlap or otherwise interfere with HRA's duties under this Agreement during the term of this Agreement from the EFFECTIVE Date to the expiration date, and

(b) THE INVENTOR agrees that all inquiries received by THE INVENTOR from or through persons other than HRA shall be referred to or notified to HRA and HRA shall be entitled to the same share thereon as on contracts obtained by HRA.

2.3 Remuneration

(a) In consideration for its service to THE INVENTOR under the terms of this Agreement, HRA shall be entitled to a share on any royalties that are paid or become due to THE INVENTOR on the following scale:

> (i) Fifty percent (50%) on the first $100,000 of royalties accrued in any calendar year;
> (ii) Forty percent (40%) on the remainder of royalties accrued in the calendar year.

(b) THE INVENTOR agrees that all royalty payments are to be made directly to HRA. Upon receiving a royalty payment, HRA shall deduct its share and forward to THE INVENTOR the remainder of the funds received. HRA will make all records pertaining to the receipt of royalties on behalf of THE INVENTOR available to THE INVENTOR for inspection and audit purposes upon reasonable notice.

2.4 Confidentiality

THE INVENTOR acknowledges the likelihood that HRA will have to disclose THE INVENTION to potential licensees and possibly others in the performance of HRA's duties under this Agreement, and that it is likely that many such parties would refuse to accept such information in confidence. Accordingly, THE INVENTOR hereby agrees to release HRA from any previous obligations to maintain information pertaining to THE INVENTION in confidence.

Notice: Intellectual Property Rights may be forfeited as a result of non-confidential disclosure of THE INVENTION. HRA encourages THE INVENTOR to explore the possibility of filing patent application(s), provisional patent application(s) or other appropriate protections for THE INVENTION prior to any loss of rights. At THE INVENTOR's request HRA will provide THE INVENTOR with the name(s) of one or more registered attorneys, which THE INVENTOR will be free to contact at THE INVENTOR's expense.

2.5 Improvements

THE INVENTOR shall own any improvements that are made by HRA in the performance of its obligation under this Agreement.

2.6 Models and Exhibits

HRA shall own any models or exhibits that may be prepared for or by HRA in the performance of its obligations under this Agreement.

2.7 Term and Cancellation

Either party may terminate this Agreement without cause upon thirty (30) days written notice, although THE INVENTOR may not so terminate this Agreement within the first six (6) months of the effective date of the Agreement. In the event that either party so Terminates this Agreement and later enters into an agreement with a business or individual first contacted by HRA, a share shall be payable to HRA according to the provisions of Section 2.3 (a) and (b) above as if the Agreement is still in force.

2.8 Liability

THE INVENTOR shall hold HRA harmless against any liability that may arise in connection with THE INVENTION.

III. MISCELLANEOUS

3.1 Severability. If any provision of this Agreement is or becomes or is deemed invalid, illegal or unenforceable in any jurisdiction, such provision shall be deemed amended to conform to applicable laws so as to be valid and enforceable or, if it cannot be so amended without materially altering the intention of the parties, it shall be stricken and the remainder of this Agreement shall remain in full force and effect.

3.2 Governing Law and Jurisdiction. This Agreement shall be deemed to have been entered into, and shall be construed and enforced in accordance with the laws of the United States of America and the State of Pennsylvania. Any disputes involving this Agreement that for any reason are not subject to arbitration as provided in Section 3.7 herein shall be sited in state or federal court that is located in Philadelphia County, Pennsylvania.

3.3 *Waiver.* No waiver of any right under this Agreement shall be in effect unless contained in writing signed by the party charged with such waiver and no waiver of any right arising from any breach or failure to perform shall be deemed to be a waiver of any future breach or failure of any other right arising under this Agreement.

3.4 *Headings.* Such headings contained herein are included for convenience only and form no part of the Agreement.

3.5 *Costs.* In the event of any controversy, claim or dispute between the parties hereto arising out of or relating to this Agreement or the terms thereof, the prevailing party shall be entitled to recover from the losing party reasonable attorney's fees and reasonable costs.

3.6 *Integration Amendment.* This Agreement constitutes the entire agreement between the parties hereto with respect to the subject matter thereof and supersedes and voids any and all prior agreements, understandings, promises and representations made by either party to the other concerning the subject matter hereof and the terms applicable hereof. This Agreement may not be released, discharged, amended or modified in any manner except by an instrument in writing signed by duly authorized representatives of the parties hereto.

3.7 *Arbitration.* Any dispute relating to this Agreement shall be decided by binding arbitration by an arbitrator who is certified by the American Arbitration Association and is acceptable by both parties. The site of any arbitration shall be in Philadelphia County, Pennsylvania.

IN WITNESS HEREOF, THE INVENTOR and HRA have caused this Agreement to be duly executed on the date first written below:

THE INVENTOR: HRA:

By:_____   By:_____

Title: _____   Title:_____

                                        Execution Date: _____

PTO/SB/01 (10-01)
Approved for use through 10/31/2002. OMB 0651-0032
U.S. Patent and Trademark Office; U.S. DEPARTMENT OF COMMERCE

Under the Paperwork Reduction Act of 1995, no persons are required to respond to a collection of information unless it contains a valid OMB control number.

DECLARATION FOR UTILITY OR DESIGN PATENT APPLICATION (37 CFR 1.63)	**Attorney Docket Number**	
	First Named Inventor	
	COMPLETE IF KNOWN	
☐ Declaration Submitted with Initial Filing **OR** ☐ Declaration Submitted after Initial Filing (surcharge (37 CFR 1.16 (e)) required)	Application Number	
	Filing Date	
	Art Unit	
	Examiner Name	

As the below named inventor, I hereby declare that:

My residence, mailing address, and citizenship are as stated below next to my name.

I believe I am the original and first inventor of the subject matter which is claimed and for which a patent is sought on the invention entitled:

(Title of the Invention)

the specification of which

☐ is attached hereto

OR

☐ was filed on (MM/DD/YYYY) [   ] as United States Application Number or PCT International

Application Number [   ] and was amended on (MM/DD/YYYY) [   ] (if applicable).

I hereby state that I have reviewed and understand the contents of the above identified specification, including the claims, as amended by any amendment specifically referred to above.

I acknowledge the duty to disclose information which is material to patentability as defined in 37 CFR 1.56, including for continuation-in-part applications, material information which became available between the filing date of the prior application and the national or PCT international filing date of the continuation-in-part application.

I hereby claim foreign priority benefits under 35 U.S.C. 119(a)-(d) or (f), or 365(b) of any foreign application(s) for patent, inventor's or plant breeder's rights certificate(s), or 365(a) of any PCT international application which designated at least one country other than the United States of America, listed below and have also identified below, by checking the box, any foreign application for patent, inventor's or plant breeder's rights certificate(s), or any PCT international application having a filing date before that of the application on which priority is claimed.

Prior Foreign Application Number(s)	Country	Foreign Filing Date (MM/DD/YYYY)	Priority Not Claimed	Certified Copy Attached? YES	NO
			☐	☐	☐
			☐	☐	☐
			☐	☐	☐
			☐	☐	☐

☐ Additional foreign application numbers are listed on a supplemental priority data sheet PTO/SB/02B attached hereto:

[Page 1 of 2]

Burden Hour Statement: This form is estimated to take 21 minutes to complete. Time will vary depending upon the needs of the individual case. Any comments on the amount of time you are required to complete this form should be sent to the Chief Information Officer, U.S. Patent and Trademark Office, Washington, DC 20231. DO NOT SEND FEES OR COMPLETED FORMS TO THIS ADDRESS. SEND TO: Assistant Commissioner for Patents, Washington, DC 20231.

PTO/SB/01A (10-01)
Approved for use through 10/31/2002. OMB 0651-0032
U.S. Patent and Trademark Office; U.S. DEPARTMENT OF COMMERCE
Under the Paperwork Reduction Act of 1995, no persons are required to respond to a collection of information unless it displays a valid OMB control number.

DECLARATION (37 CFR 1.63) FOR UTILITY OR DESIGN APPLICATION USING AN APPLICATION DATA SHEET (37 CFR 1.76)

Title of Invention	

As the below named inventor(s), I/we declare that:

This declaration is directed to:

☐ The attached application, or

☐ Application No. _____, filed on_____,

☐ as amended on _____(if applicable);

I/we believe that I/we am/are the original and first inventor(s) of the subject matter which is claimed and for which a patent is sought;

I/ we have reviewed and understand the contents of the above-identified application, including the claims, as amended by any amendment specifically referred to above;

I/we acknowledge the duty to disclose to the United States Patent and Trademark Office all information known to me/us to be material to patentability as defined in 37 CFR 1.56, including for continuation-in-part applications, material information which became available between the filing date of the prior application and the national or PCT International filing date of the continuation-in-part application.

All statements made herein of my/own knowledge are true, all statements made herein on information and belief are believed to be true, and further that these statements were made with the knowledge that willful false statements and the like are punishable by fine or imprisonment, or both, under 18 U.S.C. 1001, and may jeopardize the validity of the application or any patent issuing thereon.

FULL NAME OF INVENTOR(S)

Inventor one: _____

Signature: _____ Citizen of: _____

Inventor two: _____

Signature: _____ Citizen of: _____

Inventor three: _____

Signature: _____ Citizen of: _____

Inventor four: _____

Signature: _____ Citizen of: _____

☐ Additional inventors are being named on _____additional form(s) attached hereto.

Burden Hour Statement: This collection of information is required by 35 U.S.C. 115 and 37 CFR 1.63. The information is used by the public to file (and the USPTO to process) an application. Confidentiality is governed by 35 U.S.C. 122 and 37 CFR 1.14. This form is estimated to take 1 minute to complete. This time will vary depending upon the needs of the individual case. Any comments on the amount of time you are required to complete this form should be sent to the Chief Information Officer, U.S. Patent and Trademark Office, Washington, DC 20231. DO NOT SEND FEES OR COMPLETED FORMS TO THIS ADDRESS. SEND TO: Assistant Commissioner for Patents, Washington, DC 20231.

PTO/SB/01 (10-01)
Approved for use through 10/31/2002. OMB 0651-0032
U.S. Patent and Trademark Office; U.S. DEPARTMENT OF COMMERCE
Under the Paperwork Reduction Act of 1995, no persons are required to respond to a collection of information unless it contains a valid OMB control number.

DECLARATION — Utility or Design Patent Application

Direct all correspondence to: ☐	Customer Number or Bar Code Label		OR ☐	Correspondence address below

Name

Address

City		State	ZIP

Country	Telephone	Fax

I hereby declare that all statements made herein of my own knowledge are true and that all statements made on information and belief are believed to be true; and further that these statements were made with the knowledge that willful false statements and the like so made are punishable by fine or imprisonment, or both, under 18 U.S.C. 1001 and that such willful false statements may jeopardize the validity of the application or any patent issued thereon.

NAME OF SOLE OR FIRST INVENTOR :	☐ A petition has been filed for this unsigned inventor

Given Name (first and middle [if any])	**Family Name** or Surname

Inventor's Signature	**Date**

Residence: City	**State**	**Country**	**Citizenship**

Mailing Address

City	**State**	**ZIP**	**Country**

NAME OF SECOND INVENTOR:	☐ A petition has been filed for this unsigned inventor

Given Name (first and middle [if any])	**Family Name** or Surname

Inventor's Signature	**Date**

Residence: City	**State**	**Country**	**Citizenship**

Mailing Address

City	**State**	**ZIP**	**Country**

☐ Additional inventors are being named on the ____supplemental Additional Inventor(s) sheet(s) PTO/SB/02A attached hereto.

[Page 2 of 2]

PTO/SB/05 (03-01)
Approved for use through 10/31/2002. OMB 0651-0032
U.S. Patent and Trademark Office; U.S. DEPARTMENT OF COMMERCE
Under the Paperwork Reduction Act of 1995, no persons are required to respond to a collection of information unless it displays a valid OMB control number.

UTILITY PATENT APPLICATION TRANSMITTAL

(Only for new nonprovisional applications under 37 CFR 1.53(b))

Attorney Docket No.	
First Inventor	
Title	
Express Mail Label No.	

APPLICATION ELEMENTS

See MPEP chapter 600 concerning utility patent application contents.

1. ☐ Fee Transmittal Form (e.g., PTO/SB/17)
   (Submit an original and a duplicate for fee processing)

2. ☐ Applicant claims small entity status.
   See 37 CFR 1.27.

3. ☐ Specification   [*Total Pages* ☐]
   (preferred arrangement set forth below)
   - Descriptive title of the invention
   - Cross Reference to Related Applications
   - Statement Regarding Fed sponsored R & D
   - Reference to sequence listing, a table,
     or a computer program listing appendix
   - Background of the Invention
   - Brief Summary of the Invention
   - Brief Description of the Drawings *(if filed)*
   - Detailed Description
   - Claim(s)
   - Abstract of the Disclosure

4. ☐ Drawing(s) *(35 U.S.C. 113)*   [*Total Sheets* ☐]

5. Oath or Declaration   [*Total Pages* ☐]

   a. ☐ Newly executed (original or copy)

   b. ☐ Copy from a prior application (37 CFR 1.63 (d))
      (for continuation/divisional with Box 18 completed)

      i. ☐ **DELETION OF INVENTOR(S)**
         Signed statement attached deleting inventor(s)
         named in the prior application, see 37 CFR
         1.63(d)(2) and 1.33(b).

6. ☐ Application Data Sheet. See 37 CFR 1.76

ADDRESS TO: Assistant Commissioner for Patents
Box Patent Application
Washington, DC 20231

7. ☐ CD-ROM or CD-R in duplicate, large table or
   Computer Program (*Appendix*)

8. Nucleotide and/or Amino Acid Sequence Submission
   (if applicable, all necessary)

   a. ☐ Computer Readable Form (CRF)

   b. Specification Sequence Listing on:

      i. ☐ CD-ROM or CD-R (2 copies); or

      ii. ☐ paper

   c. ☐ Statements verifying identity of above copies

ACCOMPANYING APPLICATION PARTS

9. ☐ Assignment Papers (cover sheet & document(s))

10. ☐ 37 CFR 3.73(b) Statement   ☐ Power of
    (when there is an assignee)   Attorney

11. ☐ English Translation Document *(if applicable)*

12. ☐ Information Disclosure   ☐ Copies of IDS
    Statement (IDS)/PTO-1449   Citations

13. ☐ Preliminary Amendment

14. ☐ Return Receipt Postcard (MPEP 503)
    (Should be specifically itemized)

15. ☐ Certified Copy of Priority Document(s)
    (if foreign priority is claimed)

16. ☐ Nonpublication Request under 35 U.S.C. 122
    (b)(2)(B)(i). Applicant must attach form PTO/SB/35
    or its equivalent.

17. ☐ Other: ....
    ....

18. If a CONTINUING APPLICATION, *check appropriate box, and supply the requisite information below and in a preliminary amendment,
or in an Application Data Sheet under 37 CFR 1.76:*

☐ Continuation   ☐ Divisional   ☐ Continuation-in-part (CIP)   of prior application No.:_____/_____

Prior application information: Examiner:_____   Group Art Unit: _____

**For CONTINUATION OR DIVISIONAL APPS only: The entire disclosure of the prior application, from which an oath or declaration is supplied under
Box 5b, is considered a part of the disclosure of the accompanying continuation or divisional application and is hereby incorporated by reference.
The incorporation <u>can only</u> be relied upon when a portion has been inadvertently omitted from the submitted application parts.**

19. CORRESPONDENCE ADDRESS

☐ *Customer Number or Bar Code Label*   *(Insert customer No. or Attach bar code label here)*   or ☐ *Correspondence address below*

Name	
Address	
City		State		Zip Code	
Country		Telephone		Fax	

Name (Print/Type)		Registration No. (Attorney/Agent)	
Signature		Date	

PTO/SB/16 (10-01)
Approved for use through10/31/2002. OMB 0651-0032
U.S. Patent and Trademark Office; U.S. DEPARTMENT OF COMMERCE
Under the Paperwork Reduction Act of 1995, no persons are required to respond to a collection of information unless it displays a valid OMB control number.

PROVISIONAL APPLICATION FOR PATENT COVER SHEET
This is a request for filing a PROVISIONAL APPLICATION FOR PATENT under 37 CFR 1.53(c).

Express Mail Label No.

INVENTOR(S)

Given Name (first and middle [if any])	Family Name or Surname	Residence (City and either State or Foreign Country)

☐ *Additional inventors are being named on the _____ separately numbered sheets attached hereto*

TITLE OF THE INVENTION (500 characters max)

CORRESPONDENCE ADDRESS

Direct all correspondence to:

☐ Customer Number

Type Customer Number here   →   Place Customer Number Bar Code Label here

OR

☐ Firm *or* Individual Name

Address

Address

City		State		ZIP	
Country		Telephone		Fax	

ENCLOSED APPLICATION PARTS (check all that apply)

☐ Specification   *Number of Pages*

☐ Drawing(s)   *Number of Sheets*

☐ Application Data Sheet. See 37 CFR 1.76

☐ CD(s), Number

☐ Other (specify)

METHOD OF PAYMENT OF FILING FEES FOR THIS PROVISIONAL APPLICATION FOR PATENT

☐ Applicant claims small entity status. See 37 CFR 1.27.

☐ A check or money order is enclosed to cover the filing fees

☐ The Commissioner is hereby authorized to charge filing fees or credit any overpayment to Deposit Account Number:

☐ Payment by credit card. Form PTO-2038 is attached.

FILING FEE AMOUNT ($)

The invention was made by an agency of the United States Government or under a contract with an agency of the United States Government.

☐ No.

☐ Yes, the name of the U.S. Government agency and the Government contract number are: _____

_____

Respectfully submitted,

SIGNATURE _____

TYPED or PRINTED NAME _____

TELEPHONE _____

Date

REGISTRATION NO.
(if appropriate)
Docket Number:

USE ONLY FOR FILING A PROVISIONAL APPLICATION FOR PATENT

PROVISIONAL APPLICATION COVER SHEET
Additional Page

PTO/SB/16 (10-01)
Approved for use through 10/31/2002. OMB 0651-0032
U.S. Patent and Trademark Office; U.S. DEPARTMENT OF COMMERCE
Under the Paperwork Reduction Act of 1995, no persons are required to respond to a collection of information unless it displays a valid OMB control number.

Docket Number	

INVENTOR(S)/APPLICANT(S)		
Given Name (first and middle [if any])	Family or Surname	Residence (City and either State or Foreign Country)

Number _____ of _____

WARNING: Information on this form may become public. Credit card information should not be included on this form. Provide credit card information and authorization on PTO-2038.

PTO/SB/95 (08-00)
Approved for use through 05/31/2002. OMB 0651-0030
U.S. Patent and Trademark Office; U.S. DEPARTMENT OF COMMERCE
Under the Paperwork Reduction Act of 1995, no persons are required to respond to a collection of information unless it displays a valid OMB control number.

Disclosure Document Deposit Request

Mail to:

    Box DD
    Assistant Commissioner for Patents
    Washington, DC 20231

Inventor(s): _____

Title of Invention: _____

Enclosed is a disclosure of the above-titled invention consisting of _____sheets of description and _____sheets of drawings. A check or money order in the amount of _____is enclosed to cover the fee (37 CFR 1.21(c)).

The undersigned, being a named inventor of the disclosed invention, requests that the enclosed papers be accepted under the Disclosure Document Program, and that they be preserved for a period of two years.

_____	_____
Signature of Inventor	Address
_____	_____
Typed of printed name	
_____	_____
Date	City, State, Zip

NOTICE OF INVENTORS

It should be clearly understood that a Disclosure Document is not a patent application, nor will its receipt date in any way become the effective filing date of a later filed patent application. A Disclosure Document may be relied upon only as evidence of conception of an invention and a patent application should be diligently filed if patent protection is desired.

Your Disclosure Document will be retained for two years after the date it was received by the United States Patent and Trademark Office (USPTO) and will be destroyed thereafter unless it is referred to in a related patent application filed within the two-year period. The Disclosure Document may be referred to by way of a letter of transmittal in a new patent application or by a separate letter filed in a pending application. Unless it is desired to have the USPTO retain the Disclosure Document beyond the two-year period, it is not required that it be referred to in the patent application.

The two-year retention period should not be considered to be a "grace period" during which the inventor can wait to file his/her patent application without possible loss of benefits. It must be recognized that in establishing priority of invention an affidavit or testimony referring to a Disclosure Document must usually also establish diligence in completing the invention or in filing the patent application since the filing of the Disclosure Document.

If you are not familiar with what is considered to be "diligence in completing the invention" or "reduction to practice" under the patent law or if you have other questions about patent matters, you are advised to consult with an attorney or agent registered to practice before the USPTO. The publication, *Attorneys and Agents Registered to Practice Before the United States Patent and Trademark Office,* is available from the **Superintendent of Documents, Washington, DC 20402.** Patent attorneys and agents are also listed in the telephone directory of most major cities. Also, many large cities have associations of patent attorneys which may be consulted.

You are also reminded that any public use or sale in the United States or publication of your invention anywhere in the world more than one year prior to the filing of a patent application on that invention will prohibit the granting of a patent on it.

Disclosures of inventions which have been understood and witnessed by persons and/or notarized are other examples of evidence which may also be used to establish priority.

There is a nationwide network of Patent and Trademark Depository Libraries (PTDLs), which have collections of patents and patent-related reference materials available to the public, including automated access to USPTO databases. Publications such as *General Information Concerning Patents* are available at the PTDLs, as well as the USPTO's Web site at www.uspto.gov. To find out the location of the PTDL closest to you, please consult the complete listing of all PTDLs that appears on the USPTO's Web site or in every issue of the Official Gazette, or call the USPTO's General Information Services at 800-PTO-9199 (800-786-9199) or 703-308-HELP (703-308-4357). To insure assistance from a PTDL staff member, you may wish to contact a PTDL prior to visiting to learn about its collections, services, and hours.

Burden Hour Statement: This collection of information is used by the public to file (and by the USPTO to process) Disclosure Document Deposit Requests. Confidentiality is governed by 35 USC 122 and 37 CFR 1.14. This collection is estimated to take 12 minutes to complete, including gathering, preparing, and submitting the completed Disclosure Document Deposit Request to the USPTO. Time will vary depending upon the individual case. Any comments on the amount of time you require to complete this form and/or suggestions for reducing this burden, should be sent to the Chief Information Officer, U.S. Patent and Trademark Office, U.S. Department of Commerce, Washington, DC 20231. DO NOT SEND FEES OR COMPLETED FORMS TO THIS ADDRESS. SEND TO: Assistant Commissioner for Patents, Washington, DC 20231.

Trademark/Service Mark Application, Principle Register with Declaration

Applicant Information		
Please use the <u>Wizard</u> if there are multiple applicants.		
* <u>Name</u>		
	[If an individual, use following format: Last Name, First Name, Middle Initial/Name]	
<u>Entity Type</u>: Click on the **one** appropriate circle to indicate the applicant's entity type and enter the corresponding information.		
○ **Individual**	**Country of Citizenship**	
○ **Corporation**	**State or Country of Incorporation**	
○ **Partnership**	**State or Country Where Organized**	
	Name and Citizenship of all General Partners	
○ **Other**	**Specify Entity Type**	
	State or Country Where Organized	
* **Address**	* **Street Address**	
	* **City**	
	State	Select State ⇕
		If not listed above, please select 'OTHER' and specify here:

	* Country	**Select Country** ⬍
		If not listed above, please select 'OTHER' and specify here:
	Zip/Postal Code	
Phone Number		
Fax Number		
Internet E-Mail Address	☐ Check here to authorize the USPTO to communicate with the applicant or its representative via e-mail. NOTE: While the application may list an e-mail address for the applicant, applicant's attorney, and/or applicant's domestic representative, **only one** e-mail address may be used for correspondence, in accordance with Office policy. The applicant must keep this address current in the Office's records.	

Mark Information

Before the USPTO can register your mark, we must know exactly what it is. You can display a mark in one of two formats:
(1) typed; or (2) stylized or design. When you click on one of the two circles below, and follow the relevant instructions, the program will create a separate page that displays your mark.

WARNING: AFTER SEARCHING THE USPTO DATABASE, EVEN IF YOU THINK THE RESULTS ARE "O.K.," DO NOT ASSUME THAT YOUR MARK CAN BE REGISTERED AT THE USPTO. AFTER YOU FILE AN APPLICATION, THE USPTO MUST DO ITS OWN SEARCH AND OTHER REVIEW, AND MIGHT REFUSE TO REGISTER YOUR MARK.

	○ **Typed Format**	Click on this circle if you wish to register a word(s), letter(s), and/or number(s) in a format that can be reproduced using a typewriter. Also, only the following common punctuation marks and symbols are acceptable in a typed drawing (any other symbol requires a stylized format): . ? " - ; ( ) % $ @ + , ! ' : / & # * = [ ] Enter the mark here: NOTE: The mark **must** be entered in ALL upper case letters, regardless of how you actually use the mark. E.g., MONEYWISE, **not** MoneyWise.
* Mark	○ **Stylized or Design Format**	Click on this circle if you wish to register a stylized word(s), letter(s), number(s), and/or a design. The design may also include words. Click on the 'Browse' button to select GIF or JPG image file from your local drive that shows the complete, overall mark (i.e., the stylized representation of the words, e.g., or if a design that also includes words, the image of the "composite" mark, NOT just the design element). Do NOT submit a color image. [Browse...] For a stylized word(s) or letter(s), or a design that also includes a word(s), enter the LITERAL element only of the mark here:

This section is for the entry of various statements that may pertain to the mark. In no case must you enter any of these statements for the application to be accepted for filing (although you may be required to add a statement(s) to the record during the actual prosecution of the application). To select a statement, check the box and enter the specific information relevant to your mark. The following are the texts of the most commonly asserted statements:

☐ **DISCLAIMER:** "No claim is made to the exclusive right to use ⬚ apart from the mark as shown."

☐ **STIPPLING AS A FEATURE OF THE MARK:** "The stippling is a feature of the mark."

☐ **STIPPLING FOR SHADING:** "The stippling is for shading purposes only."

☐ **PRIOR REGISTRATION(S):** "Applicant claims ownership of U.S. Registration Number(s) ⬚."

☐ **DESCRIPTION OF THE MARK:** "The mark consists of ⬚."

☐ **TRANSLATION:** "The foreign wording in the mark translates into English as ⬚."

☐ **TRANSLITERATION:** "The non-Latin character(s) in the mark transliterate into ⬚, and this means ⬚ in English."

☐ **§2(f), based on Use:** "The mark has become distinctive of the goods/services through the applicant's substantially exclusive and continuous use in commerce for at least the five years immediately before the date of this statement."

☐ **§2(f), based on Prior Registration(s):** "The mark has become distinctive of the goods/services as evidenced by the ownership on the Principal Register for the same mark for related goods or services of U.S. Registration No(s). ⬚."

☐ **§2(f), IN PART, based on Use:** "⬚ has become distinctive of the goods/services through the applicant's substantially exclusive and continuous use in commerce for at least the five years immediately before the date of this statement."

☐ **§2(f), IN PART, based on Prior Registration(s):** "⬚ has become distinctive of the goods/services as evidenced by the ownership on the Principal Register for the same mark for related goods or services of U.S. Registration No(s). ⬚."

☐ **NAME(S), PORTRAIT(S), SIGNATURE(S) OF INDIVIDUAL(S):**

○ "The name(s), portrait(s), and/or signature(s) shown in the mark identifies ⬚, whose consent(s) to register will be submitted.

○ "The name(s), portrait(s), and/or signature(s) shown in the mark does not identify a particular living individual.

☐ **USE OF THE MARK IN ANOTHER FORM:** "The mark was first used anywhere in a different form other than that sought to be registered on ⬚, and in commerce on ⬚."

☐ **CONCURRENT USE:** Enter the appropriate concurrent use information, e.g., specify the goods and the geographic area for which registration is sought.

Additional Statement

BASIS FOR FILING AND GOODS AND/OR SERVICES INFORMATION

Applicant requests registration of the trademark/service mark identified above with the Patent and Trademark Office on the Principal Register established by the Act of July 5, 1946 (15 U.S.C. §1051 et seq.) for the following Class(es) and Goods and/or Services, and checks the basis that covers those specific Goods or Services. More than one basis may be selected, but do **NOT** claim both §§1(a) and 1(b) for the identical goods or services in one application.

☐ **Section 1(a), Use in Commerce: Applicant is using or is using through a related company the mark in commerce on or in connection with the below identified goods and/or services. 15 U.S.C. § 1051(a), as amended. Applicant attaches one specimen for *each class* showing the mark as used in commerce on or in connection with any item in the class of listed goods and/or services. If filing electronically, applicant must attach a JPG or GIF specimen image file for each international class, regardless of whether the mark itself is in a typed drawing format or is in a stylized format or a design. Unlike the mark image file, a specimen image file may be in color (i.e., if color is being claimed as a feature of the mark, then the specimen image should show use of the actual color(s) claimed).**

Specimen Image File

Click on the 'Browse' button to select GIF or JPG image file that contains the specimen from applicant's local drive.

[   ] [Browse...]

Describe what the specimen submitted consists of:

[   ]

International Class	[ ] If known, enter class number 001 - 045, A, B, or 200
* **Listing of Goods and/or Services** *USPTO Goods/ Services Manual*	
Date of First Use of Mark Anywhere	at least as early as: [ ] MM/DD/YYYY
Date of First Use of the Mark in Commerce	[ ] MM/DD/YYYY

☐ **Section 1(b), Intent to Use:** Applicant has a bona fide intention to use or use through a related company the mark in commerce on or in connection with the goods and/or services identified below (15 U.S.C. §1051(b)).

International Class	[ ] If known, enter class number 001 - 045, A, B, or 200
* **Listing of Goods and/or Services** *USPTO Goods/ Services Manual*	

☐ **Section 44(d), Priority based on foreign filing:** Applicant has a bona fide intention to use the mark in commerce on or in connection with the goods and/or services identified below, and asserts a claim of priority based upon a foreign application in accordance with 15 U.S.C. §1126(d).

International Class	[ ] If known, enter class number 001 - 045, A, B, or 200
* **Listing of Goods and/or Services** *USPTO Goods/ Services Manual*	
Country of Foreign Filing	[Select Country ▼] If not listed above, please select 'OTHER' and specify here: [ ]
Foreign Application Number	[ ] NOTE:If possible, enter no more than 12 characters. Eliminate all spaces and non-alphanumeric characters.

	Date of Foreign Filing	MM/DD/YYYY

☐ **Section 44(e), Based on Foreign Registration:** Applicant has a bona fide intention to use the mark in commerce on or in connection with the above identified goods and/or services, and will submit a certification or certified copy of the foreign registration before the application may proceed to registration, in accordance with 15 U. S.C. 1126(e), as amended.

International Class	If known, enter class number 001 - 045, A, B, or 200
* Listing of Goods and/or Services *USPTO Goods/ Services Manual*	
Country of Foreign Registration	Select Country ⬍ If not listed above, please select 'OTHER' and specify here:
Foreign Registration Number	NOTE:If possible, enter no more than 12 characters. Eliminate all spaces and non-alphanumeric characters.
Foreign Registration Date	MM/DD/YYYY
Renewal Date for Foreign Registration	MM/DD/YYYY
Expiration Date of Foreign Registration	MM/DD/YYYY

☐ Check here if an <u>attorney</u> is filing this application on behalf of applicant(s). Otherwise, click on <u>Domestic Representative</u> to continue.

Attorney Information

Correspondent Attorney Name		
Individual Attorney Docket/Reference Number		
Other Appointed Attorney(s)		
Attorney Address	**Street Address**	
	City	
	State	Select State ⬍ If not listed above, please select 'OTHER' and specify here:
	Country	Select Country ⬍ If not listed above, please select 'OTHER' and specify here:
	Zip/Postal Code	
Firm Name		
Phone Number		
FAX Number		
Internet E-Mail Address		☐ Check here to <u>authorize</u> the USPTO to communicate with the applicant or its representative via e-mail. NOTE: While the application may list an e-mail address for the applicant, applicant's attorney, and/or applicant's domestic representative, **only one** e-mail address may be used for correspondence, in accordance with Office <u>policy</u>. The applicant must keep this address current in the Office's records.

☐ Check here if the applicant has appointed a <u>Domestic Representative</u>. **A Domestic Representative is REQUIRED if the applicant's address is outside the United States.** Otherwise, click on <u>Fee Information</u> to continue.

Domestic Representative

The applicant **must** appoint a Domestic Representative if the applicant's address is outside the United States. The following is hereby appointed applicant's representative upon whom notice or process in the proceedings affecting the mark may be served.

Representative's Name		
Address	**Street Address**	
	City	
	State	Select State ⬍ If not listed above, please select 'OTHER' and specify here:
	Zip Code	
Firm Name		
Phone Number		
FAX Number		
Internet E-Mail Address		☐ Check here to authorize the USPTO to communicate with the applicant or its representative via e-mail. NOTE: While the application may list an e-mail address for the applicant, applicant's attorney, and/or applicant's domestic representative, **only one** e-mail address may be used for correspondence, in accordance with Office policy. The applicant must keep this address current in the Office's records.

Fee Information

Number of Classes Paid   1 ⬍

Note: The total fee is computed based on the Number of Classes in which the goods and/or services associated with the mark are classified.

$ [325] = **Number of Classes Paid x $325 (per class)**

* **Amount** $ [   ]

NOTE: TEAS has changed its payment options and procedures. Three options (credit card, automated deposit account (New!), and Electronic Funds Transfer (New!)) will now appear after clicking on the PAY/SUBMIT button, which is available on the bottom of the Validation Page after completing and validating the application form.

Declaration

The undersigned, being hereby warned that willful false statements and the like so made are punishable by fine or imprisonment, or both, under 18 U.S.C. §1001, and that such willful false statements may jeopardize the validity of the application or any resulting registration, declares that he/she is properly authorized to execute this application on behalf of the applicant; he/she believes the applicant to be the owner of the trademark/service mark sought to be registered, or, if the application is being filed under 15 U.S.C. §1051(b), he/she believes applicant to be entitled to use such mark in commerce; to the best of his/her knowledge and belief no other person, firm, corporation, or association has the right to use the mark in commerce, either in the identical form thereof or in such near resemblance thereto as to be likely, when used on or in connection with the goods/services of such other person, to cause confusion, or to cause mistake, or to deceive; and that all statements made of his/her own knowledge are true; and that all statements made on information and belief are believed to be true.

Electronic Signature

The application will not be "signed" in the sense of a traditional paper document. To verify the contents of the application, the signatory must enter any combination of alpha/numeric characters that has been specifically adopted to serve the function of the signature, preceded and followed by the forward slash (/) symbol. Acceptable "signatures" could include: /john doe/; /jd/; and /123-4567/. The application may still be verified to check for missing information or errors even if the signature and date signed fields are left blank.

Signature [   ]   Date Signed [   ]
                                                     MM/DD/YYYY
Signatory's Name [   ]

Signatory's Position [   ]

Click on the desired action:

The "Validate Form" function allows you to run an automated check to ensure that all mandatory fields have been completed. You will receive an "error" message if you have not filled in one of the five (5) fields that are considered "minimum filing requirements" under the Trademark Law Treaty Implementation Act of 1998. For other fields that the USPTO believes are important, but not mandatory, you will receive a "warning" message if the field is left blank. This warning is a courtesy, if non-completion was merely an oversight. If you so choose, you may by-pass that "warning" message and validate the form (however, you cannot by-pass an "error" message).

[Validate Form]   [Reset Form]

Note: To either print the completed application, in whole or in part, download and save the validated application, or electronically submit the application to the USPTO, click on the Validate Form button.

Privacy Policy Statement

FEE CHANGES

Fees are effective through June 30, 2002. After that date, check the Copyright Office Website at www.loc.gov/copyright or call (202) 707-3000 for current fee information.

FORM TX

For a Nondramatic Literary Work
UNITED STATES COPYRIGHT OFFICE

REGISTRATION NUMBER

_____

TX _____ TXU _____

EFFECTIVE DATE OF REGISTRATION

_____

Month   Day   Year

DO NOT WRITE ABOVE THIS LINE. IF YOU NEED MORE SPACE, USE A SEPARATE CONTINUATION SHEET.

1

TITLE OF THIS WORK ▼

PREVIOUS OR ALTERNATIVE TITLES ▼

PUBLICATION AS A CONTRIBUTION If this work was published as a contribution to a periodical, serial, or collection, give information about the collective work in which the contribution appeared.   Title of Collective Work ▼

If published in a periodical or serial give: Volume ▼   Number ▼   Issue Date ▼   On Pages ▼

2 a

NAME OF AUTHOR ▼

DATES OF BIRTH AND DEATH
Year Born ▼   Year Died ▼

Was this contribution to the work a "work made for hire"?
☐ Yes
☐ No

AUTHOR'S NATIONALITY OR DOMICILE
Name of Country
OR { Citizen of ▶ _____
Domiciled in▶ _____

WAS THIS AUTHOR'S CONTRIBUTION TO THE WORK
Anonymous?   ☐ Yes ☐ No
Pseudonymous? ☐ Yes ☐ No

If the answer to either of these questions is "Yes," see detailed instructions.

NATURE OF AUTHORSHIP Briefly describe nature of material created by this author in which copyright is claimed. ▼

NOTE

Under the law, the "author" of a "work made for hire" is generally the employer, not the employee (see instructions). For any part of this work that was "made for hire" check "Yes" in the space provided, give the employer (or other person for whom the work was prepared) as "Author" of that part, and leave the space for dates of birth and death blank.

b

NAME OF AUTHOR ▼

DATES OF BIRTH AND DEATH
Year Born ▼   Year Died ▼

Was this contribution to the work a "work made for hire"?
☐ Yes
☐ No

AUTHOR'S NATIONALITY OR DOMICILE
Name of Country
OR { Citizen of ▶ _____
Domiciled in▶ _____

WAS THIS AUTHOR'S CONTRIBUTION TO THE WORK
Anonymous?   ☐ Yes ☐ No
Pseudonymous? ☐ Yes ☐ No

If the answer to either of these questions is "Yes," see detailed instructions.

NATURE OF AUTHORSHIP Briefly describe nature of material created by this author in which copyright is claimed. ▼

c

NAME OF AUTHOR ▼

DATES OF BIRTH AND DEATH
Year Born ▼   Year Died ▼

Was this contribution to the work a "work made for hire"?
☐ Yes
☐ No

AUTHOR'S NATIONALITY OR DOMICILE
Name of Country
OR { Citizen of ▶ _____
Domiciled in▶ _____

WAS THIS AUTHOR'S CONTRIBUTION TO THE WORK
Anonymous?   ☐ Yes ☐ No
Pseudonymous? ☐ Yes ☐ No

If the answer to either of these questions is "Yes," see detailed instructions.

NATURE OF AUTHORSHIP Briefly describe nature of material created by this author in which copyright is claimed. ▼

3 a

YEAR IN WHICH CREATION OF THIS WORK WAS COMPLETED This information must be given ◀Year in all cases.

b DATE AND NATION OF FIRST PUBLICATION OF THIS PARTICULAR WORK
Complete this information ONLY if this work has been published.
Month▶ _____ Day▶ _____ Year▶ _____
◀ Nation

4

COPYRIGHT CLAIMANT(S) Name and address must be given even if the claimant is the same as the author given in space 2. ▼

TRANSFER If the claimant(s) named here in space 4 is (are) different from the author(s) named in space 2, give a brief statement of how the claimant(s) obtained ownership of the copyright. ▼

See instructions before completing this space.

APPLICATION RECEIVED

ONE DEPOSIT RECEIVED

TWO DEPOSITS RECEIVED

FUNDS RECEIVED

DO NOT WRITE HERE
OFFICE USE ONLY

MORE ON BACK ▶ • Complete all applicable spaces (numbers 5-9) on the reverse side of this page.
• See detailed instructions.   • Sign the form at line 8.

DO NOT WRITE HERE

Page 1 of _____ pages

EXAMINED BY		FORM TX
CHECKED BY		
☐ CORRESPONDENCE Yes		FOR COPYRIGHT OFFICE USE ONLY

DO NOT WRITE ABOVE THIS LINE. IF YOU NEED MORE SPACE, USE A SEPARATE CONTINUATION SHEET.

PREVIOUS REGISTRATION Has registration for this work, or for an earlier version of this work, already been made in the Copyright Office?

☐ Yes ☐ No If your answer is "Yes," why is another registration being sought? (Check appropriate box.) ▼

a. ☐ This is the first published edition of a work previously registered in unpublished form.

b. ☐ This is the first application submitted by this author as copyright claimant.

c. ☐ This is a changed version of the work, as shown by space 6 on this application.

If your answer is "Yes," give: **Previous Registration Number** ▶   **Year of Registration** ▶

5

DERIVATIVE WORK OR COMPILATION

Preexisting Material Identify any preexisting work or works that this work is based on or incorporates. ▼

a **6**

Material Added to This Work Give a brief, general statement of the material that has been added to this work and in which copyright is claimed. ▼

b

See instructions before completing this space.

DEPOSIT ACCOUNT If the registration fee is to be charged to a Deposit Account established in the Copyright Office, give name and number of Account.

Name ▼   **Account Number** ▼

a **7**

CORRESPONDENCE Give name and address to which correspondence about this application should be sent.   Name/Address/Apt/City/State/ZIP ▼

b

Area code and daytime telephone number ▶   Fax number ▶

Email ▶

CERTIFICATION* I, the undersigned, hereby certify that I am the

Check only one ▶
{ ☐ author
☐ other copyright claimant
☐ owner of exclusive right(s)
☐ authorized agent of _____

8

of the work identified in this application and that the statements made by me in this application are correct to the best of my knowledge.

Name of author or other copyright claimant, or owner of exclusive right(s) ▲

Typed or printed name and date ▼ If this application gives a date of publication in space 3, do not sign and submit it before that date.

Date ▶ _____

☛ Handwritten signature (X) ▼

X _____

Certificate will be mailed in window envelope to this address:	Name ▼	YOU MUST: • Complete all necessary spaces • Sign your application in space 8	**9**
	Number/Street/Apt ▼	SEND ALL 3 ELEMENTS IN THE SAME PACKAGE: 1. Application form 2. Nonrefundable filing fee in check or money order payable to *Register of Copyrights* 3. Deposit material	As of July 1, 1999, the filing fee for Form TX is $30.
	City/State/ZIP ▼	MAIL TO: Library of Congress Copyright Office 101 Independence Avenue, S.E. Washington, D.C. 20559-6000	

June 1999—200,000   ♻ PRINTED ON RECYCLED PAPER   ☆U.S. GOVERNMENT PRINTING OFFICE: 1999-454-879/49
WEB REV: June 1999

FEE CHANGES

Fees are effective through June 30, 2002. After that date, check the Copyright Office Website at www.loc.gov/copyright or call (202) 707-3000 for current fee information.

FORM VA
For a Work of the Visual Arts
UNITED STATES COPYRIGHT OFFICE

REGISTRATION NUMBER

VA	VAU

EFFECTIVE DATE OF REGISTRATION

Month	Day	Year

DO NOT WRITE ABOVE THIS LINE. IF YOU NEED MORE SPACE, USE A SEPARATE CONTINUATION SHEET.

1

TITLE OF THIS WORK ▼   **NATURE OF THIS WORK** ▼ See instructions

PREVIOUS OR ALTERNATIVE TITLES ▼

Publication as a Contribution If this work was published as a contribution to a periodical, serial, or collection, give information about the collective work in which the contribution appeared.   **Title of Collective Work** ▼

If published in a periodical or serial give: **Volume** ▼   **Number** ▼   **Issue Date** ▼   **On Pages** ▼

2

a

NAME OF AUTHOR ▼   **DATES OF BIRTH AND DEATH**
                                                 Year Born ▼   Year Died ▼

NOTE

Under the law, the "author" of a "work made for hire" is generally the employer, not the employee (see instructions). For any part of this work that was "made for hire" check "Yes" in the space provided, give the employer (or other person for whom the work was prepared) as "Author" of that part, and leave the space for dates of birth and death blank.

Was this contribution to the work a "work made for hire"?
☐ Yes
☐ No

Author's Nationality or Domicile
Name of Country
OR { Citizen of ▶
      Domiciled in ▶

Was This Author's Contribution to the Work
Anonymous?   ☐ Yes   ☐ No
Pseudonymous? ☐ Yes   ☐ No
If the answer to either of these questions is "Yes," see detailed instructions.

NATURE OF AUTHORSHIP Check appropriate box(es). **See instructions**
☐ 3-Dimensional sculpture   ☐ Map   ☐ Technical drawing
☐ 2-Dimensional artwork   ☐ Photograph   ☐ Text
☐ Reproduction of work of art ☐ Jewelry design ☐ Architectural work

b

NAME OF AUTHOR ▼   **DATES OF BIRTH AND DEATH**
                                                 Year Born ▼   Year Died ▼

Was this contribution to the work a "work made for hire"?
☐ Yes
☐ No

Author's Nationality or Domicile
Name of Country
OR { Citizen of ▶
      Domiciled in ▶

Was This Author's Contribution to the Work
Anonymous?   ☐ Yes   ☐ No
Pseudonymous? ☐ Yes   ☐ No
If the answer to either of these questions is "Yes," see detailed instructions.

NATURE OF AUTHORSHIP Check appropriate box(es). **See instructions**
☐ 3-Dimensional sculpture   ☐ Map   ☐ Technical drawing
☐ 2-Dimensional artwork   ☐ Photograph   ☐ Text
☐ Reproduction of work of art ☐ Jewelry design ☐ Architectural work

3

a **Year in Which Creation of This Work Was Completed**
        This information must be given
        ◀ Year in all cases.

b **Date and Nation of First Publication of This Particular Work**
        Complete this information   Month ▶ _____ Day ▶ _____ Year ▶ _____
        ONLY if this work
        has been published.   ◀ Nation

4

See instructions before completing this space.

COPYRIGHT CLAIMANT(S) Name and address must be given even if the claimant is the same as the author given in space 2. ▼

Transfer If the claimant(s) named here in space 4 is (are) different from the author(s) named in space 2, give a brief statement of how the claimant(s) obtained ownership of the copyright. ▼

DO NOT WRITE HERE
OFFICE USE ONLY

APPLICATION RECEIVED

ONE DEPOSIT RECEIVED

TWO DEPOSITS RECEIVED

FUNDS RECEIVED

MORE ON BACK ▶   • Complete all applicable spaces (numbers 5-9) on the reverse side of this page.
                     • See detailed instructions.   • Sign the form at line 8.

DO NOT WRITE HERE

Page 1 of _____ pages

EXAMINED BY	FORM VA
CHECKED BY	
☐ CORRESPONDENCE Yes	FOR COPYRIGHT OFFICE USE ONLY

DO NOT WRITE ABOVE THIS LINE. IF YOU NEED MORE SPACE, USE A SEPARATE CONTINUATION SHEET.

PREVIOUS REGISTRATION Has registration for this work, or for an earlier version of this work, already been made in the Copyright Office?

☐ Yes ☐ No If your answer is "Yes," why is another registration being sought? (Check appropriate box.) ▼

a. ☐ This is the first published edition of a work previously registered in unpublished form.

b. ☐ This is the first application submitted by this author as copyright claimant.

c. ☐ This is a changed version of the work, as shown by space 6 on this application.

If your answer is "Yes," give: **Previous Registration Number** ▼   **Year of Registration** ▼

5

DERIVATIVE WORK OR COMPILATION Complete both space 6a and 6b for a derivative work; complete only 6b for a compilation.

a. Preexisting Material Identify any preexisting work or works that this work is based on or incorporates. ▼

b. Material Added to This Work Give a brief, general statement of the material that has been added to this work and in which copyright is claimed. ▼

6

a

See instructions
before completing
this space.

b

DEPOSIT ACCOUNT If the registration fee is to be charged to a Deposit Account established in the Copyright Office, give name and number of Account.

Name ▼   **Account Number** ▼

7

a

CORRESPONDENCE Give name and address to which correspondence about this application should be sent.   Name/Address/Apt/City/State/ZIP ▼

b

Area code and daytime telephone number ▶ (   )   Fax number ▶ (   )

Email ▶

CERTIFICATION* I, the undersigned, hereby certify that I am the

check only one ▶ {
☐ author
☐ other copyright claimant
☐ owner of exclusive right(s)
☐ authorized agent of _____
Name of author or other copyright claimant, or owner of exclusive right(s) ▲

8

of the work identified in this application and that the statements made by me in this application are correct to the best of my knowledge.

Typed or printed name and date ▼ If this application gives a date of publication in space 3, do not sign and submit it before that date.

_____ Date ▶ _____

Handwritten signature (X) ▼

X _____

Certificate will be mailed in window envelope to this address:	Name ▼	**YOU MUST:** • Complete all necessary spaces • Sign your application in space 8	**9**
	Number/Street/Apt ▼	**SEND ALL 3 ELEMENTS** **IN THE SAME PACKAGE:** 1. Application form 2. Nonrefundable filing fee in check or money order payable to *Register of Copyrights* 3. Deposit material	As of July 1, 1999, the filing fee for Form VA is $30.
	City/State/ZIP ▼	**MAIL TO:** Library of Congress Copyright Office 101 Independence Avenue, S.E. Washington, D.C. 20559-6000	

*17 U.S.C. § 506(e): Any person who knowingly makes a false representation of a material fact in the application for copyright registration provided for by section 409, or in any written statement filed in connection with the application, shall be fined not more than $2,500.

June 1999—100,000
WEB REV: June 1999

♻ PRINTED ON RECYCLED PAPER

☆U.S. GOVERNMENT PRINTING OFFICE: 1999-454-879/71

Trademark Electronic Application System *file online*

The United States Patent and Trademark Office (USPTO) is pleased to present TEAS - the Trademark Electronic Application System. TEAS allows you to fill out a form and check it for completeness over the internet. Using e-TEAS you can then submit the form directly to the USPTO over the internet, making an official filing on-line. Or using PrinTEAS you can print out the completed form for mailing to the USPTO. It's your choice!

Important Notice

- **Eastern Time Controls Filing Date**.
- **Filing Fee and Refund Policy**.
- **Three Payment Options Available**.
- **Image Files Must Be in GIF or JPG Format**.
- **Contact Us**.

Click here for **technical information** and **e-TEAS Tutorial** (step-by-step instructions for filing your application directly over the Internet).

Click below to access correct form:

- Apply for a NEW mark
- File a PRE-registration form
    - file an Extension of time or Allegation of Use/ Statement of use
    - file a form after receiving your Notice of Allowance (NOA)

- File a POST-registration form or Renewal of an Existing Registered Mark

- Change Of Correspondence Address (use ONLY if you have a pending application or an existing registration with the Office).
- Petition Form (under development) For Petition Information Sheet, click here.

- Response to Office Action Form (under development)

PrinTEAS

Click here for **technical information** and **PrinTEAS Tutorial** (step-by-step instructions for completing your application on-line, to print and mail to USPTO).

Click below to access correct form:

- Apply for a NEW mark
- File a PRE-registration form
    - file an Extension of time or Allegation of Use/ Statement of use
    - file a form after receiving your Notice of Allowance (NOA)

- File a POST-registration form or Renewal of an Existing Registered Mark

- Change Of Correspondence Address does not exist in PrinTEAS format.

- Petition Form does not exist in PrinTEAS format.

- Response to Office Action Form does not exist in PrinTEAS format.

TEAS Policy and Technical Hints
Downloadable Blank New Application Form

Trademark Home

FAQ About Trademarks
Help Desk & Bug Report

USPTO Home

INVENTOR SUPPORT GROUPS LISTINGS COMPILED BY UNITED INVENTORS ASSOCIATION

UNITED STATES

Alaska
Anchorage
Alaska Inventors and Entrepreneurs
907-563-4337
Inventor@arctic.net

Wasilla
Inventors Institute of Alaska
907-376-5114

Alabama
Montevallo
Invent Alabama
205-663-9982

Arkansas
Dandanell
Inventors Congress, Inc.
501-229-4515

Arizona
Tucson
Inventors Association of Arizona
520-751-9966
support@azinventors.org

California
Huntington Beach
Inventors Forum
714-540-2491
www.inventorsforum.org

Manteca
Central Valley Inventors Association
209-239-5414
cdesigns@softcom.net

Mountain View
Inventors Alliance
650-964-1576
andrewinvents@onebox.com

Redding
Inventors Alliance of Northern
California
530-241-8427
sagn@awwwsome.com

San Diego
Inventors Forum of San Diego
858-451-1028
emovex@aol.com

Santa Rosa
Idea to Market Network
800-ITM-3210
sidnee@ap.net

Sebastopol
American Inventor Network
707-823-3865

Stanton
InventNET Forum
info@inventnet.com

Whittier
Inventors Forum—Whittier
562-464-0069
info@inventorsforum.org

Colorado
Denver
Rocky Mountain Inventors Congress
303-670-3760

Connecticut

Bridgeport
  Inventors Association of Connecticut
  203-866-0720
  gamebird@compuserve.com

Danbury
  Innovators Guild
  203-790-8235
  rfaulkner@snet.net

Delaware

Wilmington
  Early Stages East
  302-777-2460
  info@earlystageseast.org

District of Columbia
  Inventors Network of the Capitol Area
  703-971-7443
  raybik@aol.com

Florida

Boynton Beach
  Inventors Society of South Florida
  954-486-2426

Fort Myers
  Edison Inventors Association, Inc.
  941-275-4332
  drghn@aol.com

Indian Harbor Beach
  Space Coast Inventors Guild
  321-773-4031

Orlando
  Inventors Council of Central Florida
  407-859-4855

St. Petersburg
  Tampa Bay Inventors Council
  727-866-0669
  tbic@patent-faq.com

Georgia

Macon
  Inventors Association of Georgia
  912-474-6948
  jrmiq@mindspring.com

Iowa

Des Moines
  Drake University Inventure Program
  515-271-2655

Idaho

Shelly
  East Idaho Inventors Forum
  208-346-6763
  wordinjj@ida.net

Illinois

Chicago
  Inventors Council
  312-939-3329
  patent@donmoyer.com

Edwardsville
  Illinois Innovators and Inventors Club
  618-656-7445
  Invent@charter-il.com

Indiana

Marion
  Indiana Inventors Association
  756-674-2845
  arhumbert@bpsinet.com

Kansas

Hoisington
  Kansas Association of Inventors
  316-653-2165
  Clayton@hoinsington.com

Wichita
  Inventors Association of South Central
    Kansas
  316-721-1866
  refried@southwind.net

Kentucky

Nicholasville

Central Kentucky Inventors and
Entrepreneurs
606-885-9593
nashky@IBM.net

Louisiana

Baton Rouge

Louisiana Inventors Association
225-752-3783
info@recyclecycle.com

Maryland

Springfield (Virginia)

Inventors Network of the Capitol Area
703-971-7443
Raybik@aol.com

Massachusetts

Briar Main

Cape Cod Inventors Association
508-349-1629

Pepperal

Greater Boston Inventors Association
978-433-2397
crholt@aol.com

Springfield

Innovators Resource Network
413-259-2006
info@irnetwork.org

Worcester

Worcester Area Inventors
508-791-0226
lore@1930@aol.com

Maine

Orono

Portland Inventors Forum
207-581-1488
jsward@maine.edu

Michigan

East Lansing

Inventors Clubs of America
517-332-3561

Flint

Inventors Council of Mid-Michigan
810-232-7909
bross@flint.org

Grand Blanc

InventorEd, Inc.
810-655-8830
rjriley@inventored.org

Minnesota

Redwood Falls

Minnesota Inventors Congress
507-637-2344
mic@invent1.org

St. Francis

Society of Minnesota Inventors
763-753-2766
paulparis@uswest.net

St. Paul

Inventors Network
651-602-3175

Missouri

Hazlewood

Women's Inventor Project
314-432-1291

Kansas City

Mid-America Inventors' Association
816-254-9542

St. Louis

Inventors' Association of St. Louis
314-432-1291

Mississippi

Mississippi SBDC Inventor Assistance
662-915-5001
blantrip@olemiss.edu

Montana
Billings
    Blue Sky Inventors
    406-586-1541

Bozeman
    Montana Inventors Association
    406-586-1541

North Dakota
Jamestown
    North Dakota Inventors Congress
    701-252-4959

Nebraska
Brainard
    Lincoln Inventors Association
    402-545-2179

Nevada
Las Vegas
    Inventors Society of Southern Nevada
    702-435-7741
    inventssn@aol.com

Reno
    Nevada Inventors Association
    775-677-4824
    information@nevadainventors.org

New Jersey
Bradley Beach
    Jersey Shore Inventors Club
    732-776-8467

Livingston
    National Society of Inventors
    973-994-9282

Union
    Kean University SBDC
    908-527-2946
    m.kostak@turbo.kean.edu

Westfield
    New Jersey Entrepreneurs Forum
    908-789-3424

New Mexico
Albuquerque
    New Mexico Inventors Club
    505-266-3541

New York
Farmingdale
    NY Society of Professional Inventors
    516-798-1490
    danweiss.PE@juno.com

Farmingdale
    Long Island Forum of Technology
    631-755-3321
    porlando@lift.org

Rochester
    Inventors Society of Western New York
    716-225-6369
    jchiello@aol.com

Ohio
Cincinnati
    Inventors Council of Cincinnati
    513-772-9333

Columbus
    Inventors Network
    616-470-0144
    13832667@msn.com

Dayton
    Inventors Council of Dayton
    937-293-3073
    geopierce@earthlink.net

North Canton
    Inventors Council of Canton
    fleisherb@aol.com

Strongsville
    Inventors Connection of Greater Cleveland
    216-226-9681
    icgc@usa.com

Twinsburg
    Inventors Network of Greater Akron
    330-425-1749

Youngstown

Youngstown-Warren Inventors Association

330-744-4481

mm@cisnet.com

Oklahoma

Oklahoma City

Oklahoma Inventors Congress

405-947-5782

wbaker@tanet.net

Oregon

Medford

South Oregon Inventors Council

541-772-3478

Pennsylvania

Erie

Pennsylvania Inventors Association

dhbutler@velocity.net

Philadelphia

American Society of Inventors

215-546-6601

hskillman@ddhs.com

Puerto Rico

Porto Rico Inventors Association

787-760-5074

acuhost@novacomm-inc.com

Rhode Island

Providence

The Center for Design and Business

401-454-6108

cfaria@risd.edu

South Carolina

Easley

Carolina Inventors Council

864-859-0066

john17@home.com

Greensville

Inventors and Entrepreneurs Association of South Carolina

864-244-1045

South Dakota

Brookings

South Dakota Inventors Congress

605-688-4184

kent_rufer@sdstate.edu

Tennessee

Knoxville

Tennessee Inventors Association

423-869-8138

bealaj@aol.com

Nashville

Inventors Association of Middle Tennessee

615-269-4346

Texas

Amarillo

Amarillo Inventors Association

806-352-6085

keifer@aol.com

Houston

Houston Inventors Association

713-686-7676

kenroddy@nol.net

Plano

Texas Inventors Association

972-312-0090

mary@asktheinventors.com

San Antonio

Technology Advocates of San Antonio/ Inventors and Entrepreneurs

210-724-2545

ehopper5@hotmail.com

Vermont

Springfield
Inventors Network of Vermont
802-885-5100
comtu@turbont.net

Virginia

Charlottesville
Blue Ridge Inventors Club
804-973-3708
mac@luckycat.com

Springfield
Inventors Network of the Capitol Area
703-971-7443
raybik@aol.com

Washington

Langley
Whidbey Island Inventor Network
360-678-0269
wiin@whidbey.com

Port Townsend
Northwest Inventors Guild
360-385-7038
aero1@waypt.com

Vancouver
Inventors Network
503-239-8299

Wisconsin

Green Bay
Inventors Network of Wisconsin
920-429-0331
jhitzier@tsnnet.com

Manawa
Central Wisconsin Inventors
Association
920-596-3092
dr.heat@mailexcite.com

CANADA

British Columbia

Vancouver
British Columbia Inventors Society
604-838-9185
admin@inventors.ca

Nova Scotia

Dartmouth
Inter Atlantic Inventors Club
902-435-5218

Ontario

Cambridge
Waterloo-Wellington Inventors Club
519-653-8848
svandyk@bserv.com

Pickering
Durham East Independent Inventors'
Group
905-686-7172
gc7591@hotmail.com

Thornhill
Women Inventors' Project–Toronto
905-731-0328

Toronto
Inventors' Alliance of Canada
416-410-7792
ellwood@netcom.ca

Saskatoon-Saskatchewan
Saskatchewan Research Council
306-933-5400

U.S. PATENT AND TRADEMARK DEPOSITORY LIBRARIES

These libraries, containing current and earlier issued patents and trademarks, are open to the public free of charge. Trained technical staff is available to assist in patent searches.

Alabama
Auburn: Auburn University:
   Ralph Brown Draughon Library   334-844-1747
Birmingham: Birmingham Public Library   205-226-3620

Alaska
Anchorage: Z.J. Loussac Public Library   907-562-7323

Arizona
Tempe: Daniel E. Noble Science and Engineering Library   602-965-7010

Arkansas
Little Rock: Arkansas State Library   501-682-2053

California
Los Angeles: Los Angeles Public Library   213-228-7220
Sacramento: California State Library:
   Library-Courts Building   916-654-0069
San Diego: San Diego Public Library   619-236-5813
San Francisco: San Francisco Public Library
   and Sunnyvale Center   415-557-4500

Colorado
Denver: Denver Public Library   303-640-6220

Connecticut
Hartford: Hartford Public Library
New Haven: New Haven Public Library   203-946-8130

Delaware
Newark: University of Delaware Library   302-831-2965

District of Columbia
Washington, D.C.: Founders Library, Howard University   202-806-7252

Florida
Fort Lauderdale: Broward County Main Library   954-357-7444
Miami: Miami-Dade Public Library   305-375-2665
Orlando: University of Central Florida Libraries   407-823-2562
Tampa: Tampa Campus Library,
    University of South Florida   813-974-2726

Georgia
Atlanta: Library and Information Center,
    Georgia Institute of Technology   404-894-4508

Hawaii
Honolulu: Hawaii State Library   808-586-3477

Idaho
Moscow: University of Idaho Library   208-885-6235

Illinois
Chicago: Chicago Public Library   312-747-4450
Springfield: Illinois State Library   217-782-5659

Indiana
Indianapolis: Indianapolis–Marion County
    Public Library   317-269-1741
West Lafayette: Siegesmund Engineering Library,
    Purdue University   317-494-2972

Iowa
Des Moines: State Library of Iowa   515-281-4118

Kansas
Wichita: Ablah Library, Wichita State University   316-978-3155

Kentucky
Louisville: Louisville Free Public Library   502-574-1611

Louisiana
Baton Rouge: Troy H. Middleton Library,
    Louisiana State Univ.   504-388-8875

Maine
Orono: Raymond H. Fogler Library, University of Maine   207-581-1678

Maryland
College Park: Engineering and Physical Sciences Library,
   University of Maryland   301-405-9157

Massachusetts
Amherst: Physical Sciences and Engineering Library,
   University of Massachusetts   413-545-1370
Boston: Boston Public Library   617-536-5400

Michigan
Ann Arbor: Media Union Library,
   University of Michigan   313-647-5735
Big Rapids: Abigail S. Timme Library,
   Ferris State University   616-592-3602
Detroit: Detroit Public Library   313-833-3379

Minnesota
Minneapolis: Minneapolis Public Library
   and Information Center   612-630-6120

Mississippi
Jackson: Mississippi Library Commission   601-359-1036

Missouri
Kansas City: Linda Hall Library   816-363-4600
St. Louis: St. Louis Public Library   314-241-2288

Montana
Butte: Montana Tech of the University of Montana   406-496-4281

Nebraska
Lincoln: Engineering Library, Nebraska Hall   402-784-6500

Nevada
Reno: University Library, University of Nevada-Reno   702-784-6500

New Jersey
Newark: Newark Public Library   973-733-7779
Piscataway: Library of Science and Medicine,
   Rutgers University   732-445-2895

New Mexico

Albuquerque: Centennial Science Library,
   University of New Mexico 505-277-4412

New York

Albany: New York State Science, Industry and
   Business Library 518-474-5355
Buffalo: Buffalo and Erie County Public Library 716-858-7101
New York: Science, Industry and Business Library ... 212-592-7000
Stony Brook: Engineering Library,
   State University of New York 516-632-7148

North Carolina

Raleigh: D.H. Hill Library,
   North Carolina State University 919-515-2935

North Dakota

Grand Forks: Chester Fritz Library,
   University of North Dakota 701-777-4888

Ohio

Akron: Akron-Summit County Public Library 330-643-9075
Cincinnati: The Public Library of Cincinnati
   and Hamilton Library 513-369-6971
Cleveland: Cleveland Public Library 216-623-2870
Columbus: Ohio State University Libraries 614-292-6175
Toledo: Toledo/Lucas County Public Library 419-259-5212

Oregon

Portland: Paul L. Boley Law Library,
   Lewis and Clark College 503-768-6786

Pennsylvania

Philadelphia: The Free Library of Philadelphia 215-686-5331
Pittsburgh: The Carnegie Library of Pittsburgh 412-622-3138
State College: Patee Library,
   Pennsylvania State University 503-768-6786

Puerto Rico

Mayagüez: General Library, University of Puerto Rico ... 787-832-4040

Rhode Island

Providence: Providence Public Library 410-455-8027

South Carolina

Clemson: R.M. Cooper Library, Clemson University   864-656-3024

South Dakota

Rapid City: Devereaux Library, South Dakota School
of Mines and Technology   605-394-1275

Tennessee

Memphis: Memphis and Shelby County Public Library   901-725-8877
Nashville: Stevenson Science and Engineering Library,
Vanderbilt University   615-322-2717

Texas

Austin: McKinney Engineering Library,
University of Texas   512-495-4500
College Station: Sterling C. Evans Library,
Texas A & M University   409-845-3826
Dallas: Dallas Public Library   214-670-1468
Houston: Fondren Library, Rice University   713-527-8101
Lubbock: Texas Tech University   806-742-2282

Utah

Salt Lake City: Marriott Library, University of Utah   801-581-8394

Vermont

Burlington: Bailey/Howe Library, University of Vermont   802-656-2542

Virginia

Richmond: James Branch Cabell Library,
Virginia Community University   804-828-1104

Washington

Seattle: Engineering Library, University of Washington   206-543-0740

West Virginia

Morgantown: Evansdale Library,
University of West Virginia   304-293-2510

Wisconsin

Madison: Kurt F. Wendt Library,
University of Wisconsin   608-262-6845
Milwaukee: Milwaukee Public Library   414-286-3051

INVENTOR'S CONFIDENTIAL DISCLOSURE AGREEMENT

INFORMATION the parties (the party disclosing the CONFIDENTIAL INFORMATION and the party receiving same are hereinafter called "DISCLOSER" and "RECIPIENT", respectively) agree as follows:

1. To be protected hereunder, CONFIDENTIAL INFORMATION must be disclosed in written or graphic form conspicuously labeled with the name of the DISCLOSER as CONFIDENTIAL INFORMATION, or disclosed aurally and be documented in detail, labeled as above, and submitted by DISCLOSER in written or graphic form to RECIPIENT within twenty (20) business days thereafter.

2. RECIPIENT agrees to receive and hold all such CONFIDENTIAL INFORMATION acquired from DISCLOSER in strict confidence and to disclose same within its own organization only, and only to those of its employees who have agreed in writing (under RECIPIENT's own blanket or specified agreement form) to protect and preserve the confidentiality of such disclosures and who are designated by RECIPIENT to evaluate the CONFIDENTIAL INFORMATION for the aforementioned purposes. Without affecting the generality of the foregoing, RECIPIENT will exercise no less care to safeguard the CONFIDENTIAL INFORMATION acquired from DISCLOSER than RECIPIENT exercises in safeguarding its own confidential or proprietary information.

3. RECIPIENT agrees that it will not disclose or use CONFIDENTIAL INFORMATION acquired from DISCLOSER, in whole or in part, for any purposes other than those expressly permitted herein. Without affecting the generality of the foregoing, RECIPIENT agrees that it will not disclose any such CONFIDENTIAL INFORMATION to any third party, or use same for its own benefit or for the benefit of any third party.

4. The foregoing restrictions on RECIPIENT's disclosure and use of CONFIDENTIAL INFORMATION acquired from DISCLOSURE shall not apply to the extent of information (i) known to RECIPIENT prior to receipt from DISCLOSER (ii) of public knowledge without breach of RECIPIENT's obligation hereunder, (iii) rightfully acquired by RECIPIENT from a third party without restriction on disclosure or use, (iv) disclosed by DISCLOSER to a third party without restriction on disclosure or use, or (v) independently developed by RECIPIENT relies as relieving it of the restrictions hereunder on disclosure or use of such CONFIDENTIAL INFORMATION, and provided further that in the case of any of events (ii), (iii), (iv), and (v), the removal of restrictions shall be effective only from and after the date of occurrence of the applicable event.

5. The furnishing of CONFIDENTIAL INFORMATION hereunder shall not constitute or be construed as a grant of any express or implied license or other right, or a covenant not

to sue or forbearance from any other right of action (except as to permitted activities hereunder), by DISCLOSER to RECIPIENT under any of DISCLOSER's patents or other intellectual property rights.

6. This Agreement shall commence as of the day and year first written above and shall continue with respect to any dislosures of CONFIDENTIAL INFORMATION by DISCLOSER to RECIPIENT within twelve (12) months thereafter, at the end of which time the Agreement shall expire, unless terminated earlier by either party at any time on ten (10) days prior written notice to the other party. Upon expiration or termination of this Agreement, RECIPIENT shall immediately cease any and all disclosures or uses of CONFIDENTIAL INFORMATION acquired from DISCLOSER (except to the extent relieved from restrictions pursuant to paragraph 4 above) and at DISCLOSER's request RECIPIENT shall promptly return all written, graphic and other tangible forms of the CONFIDENTIAL INFORMATION (including notes or other writeups thereof made by RECIPIENT in connection with the disclosures by DISCLOSER) and all copies thereof made by RECIPIENT except one copy for record retention only.

7. The obligations of RECIPIENT respecting disclosure and use of CONFIDENTIAL INFORMATION acquired from DISCLOSER shall survive expiration or termination of this Agreement and shall continue for a period of three (3) years thereafter or, with respect to any applicable portion of the CONFIDENTIAL INFORMATION, until the effective date of any of the events recited in paragraph 4, whichever occurs first. After such time RECIPIENT shall be relieved of all such obligations.

8. In the event that the parties enter into a written contract concerning a business relationship of the type contemplated herein, the provisions of such contract concerning confidentiality of information shall supersede and prevail over any conflicting provisions of this Agreement.

Each party acknowledges its acceptance of this Agreement by the signature below of its authorized officer on duplicate counterparts of the Agreement, one of which fully executed counterparts is to be retained by each party.

Date: _____ Signature (YOURS) _____

Date: _____ Signature (THEIRS) _____

PATENTCAFE® MARKET ASSESSMENT FORM

Use this form to evaluate sales potential. Recognize that the competitors you list and identify in section A & B would also be the detailed potential buyers or licensees of your technology. You can not be too thorough - update this form every few months, and keep a list of thoughts and opinions about who you may contact at the right time to present your technology.

A. Product Sales Assessment (Consumer Product)

a. The typical consumer buyer of my product (male, female, age, income, etc.) is: _____
   _____

b. This customer would buy my product because: _____

c. It satisfies these customer <u>needs</u>: _____

d. It satisfies these customer <u>wants</u>: _____

e. It satisfies these customer <u>desires</u>: _____

f. The customer could chose to buy these products instead (list all): _____
   _____

g. The customer would buy mine because: _____

h. The retail price difference between my product and the competition is: $_____ (Rule of Thumb: retail price = cost of production times 4. Verify your estimated production cost and retail price)

i. The competition sells _____ units per year; I feel my product unit sales / year will be _____.

j. The total estimated annual retail sales are $ _____; total manufacturers' sales revenue is $_____ (1/2 retail).

B. Product Sales Assessment (Commercial Product)

a. The typical business buyer of my product (include S.I.C. codes) is: _____
   _____

b. The major benefit the company would receive by buying my product is: (gov't, environmental, or safety compliance, production capacity, savings, enhanced capabilities etc., be specific - this is important to the marketability): _____
   _____

c. Instead of buying my product, the business customer can (do nothing, buy a competitive product, other): _____
   _____

d. List all your competitors and competitive products, pricing & features on a separate sheet: _____

e. The problems I see in getting my product to the purchasing decision-makers at my customer companies are (include any requirement for technical salesmen, budget approval, seasonal or geographic issues, etc.): _____
   _____

f. I feel my total annual product sales could reach $_____ in the first production year, and $_____ in year #2.

C. Technology Sales, Licensing or Future Sale of My Business Assessment

In order to establish any reasonable probability of selling or licensing technology, or selling your company once it is built up successfully, or valuating the sales price, you will have to start with your hard costs. A company will initially value your idea/product/company based on how much it would cost to reproduce from scratch instead of buying your proposal. You already know that "ideas" are nearly worthless, so you have to present a "profit" story to potential buyers. The estimated product sales analysis above will be a good start from the "sales" standpoint - this section must focus on the costs associated with getting the product manufactured. The company will find many objections to buying your offering, but the more you understand the buyer's operations, the better you can address these objections and find additional sales benefits and features.

REQUIRED BUSINESS ACTIVITY	ESTIMATED TOT. COST	TOTAL SPENT
Engineering /Design from concept through prototype	$_____	$_____
Engineering/Design costs from proto to production start-up	$_____	$_____
Ongoing Engineering costs to support changes	$_____	$_____
Working Prototype/Proof of Concept costs	$_____	$_____
Production tools, jigs, special fixtures costs	$_____	$_____
Start-up production inventory (small volume pilot run)	$_____	$_____
Total Development Costs (raw cost value of your efforts)	$_____	$_____
Total Development: Concept through pilot run (Time savings value)	_____ Weeks	_____ Weeks

* This analysis will need professional input, but will provide you an understanding of why a company would want to buy your technology (save time, save learning curve, avoid infringement etc.) and what the minimum value would be to them.

Download Patentcafe's Complete Assessment Software from http://ww.ipbookstore.com.

PATENTCAFE™ INVENTOR'S SELF-ASSESSMENT FORM

Easy instructions: Answer these questions honestly to discover your Ironman Inventing strengths and weaknesses. There are no "right" answers, just an overall perspective. Try to err on the conservative side of each item since you are not pushing for a "score". Your scores identify possible problem areas that will have to either be improved, or complemented by your teammates. Patentcafe Invention Assessment and invention supplies available at **http://ww.ipbookstore.com**.

Column headings: 1 = No - Never, 2 = Probably Not, 3 = Don't Know, 4 = Maybe, 5 = Yes

1. Why do I want to develop this invention?
- a. It represents a scientific breakthrough, is important, and will benefit mankind.
- b. I really think the product will fly - all my friends agree.
- c. I always wanted to do this product - I've had this idea for a long time.
- d. I have identified a market need for the product.
- e. I really think I can get a patent on it because it is a novel idea..
- f. I am familiar with competitive products and know how to improve upon them.
- **Questions Totals:**

2. What are my capabilities and qualifications to bring this product idea to fruition?
- a. I am a qualified engineer and have developed many products.
- b. I am accountant/financial professional and have developed financial plans.
- c. I am a manufacturing professional familiar with production processes.
- d. I am a qualified team-building and project manager.
- e. I am a marketing/sales professional with experience in new product introductions.
- f. I am a qualified investor and have launched new products & businesses before.
- g. I know how to read legal contracts and understand the implications of signing them.
- **Questions Totals:**   **Total Combined Score: (add all lines) _____**

3. I know what is involved in embarking on this project because:
- a. I am formally educated in business, technology, marketing or management.
- b. I've tried to develop products before, and have failed.
- c. I've developed projects before and have succeeded.
- d. I've studied "all the books" on inventing and business start-ups.
- **Questions Totals:**   **Total Combined Score: (add all lines) _____**

4. What resources do I have to develop this project?
- a. I have access to qualified market research professionals.
- b. I have or have access to a machine shop or prototype facility,
- c. I have time in the evenings and weekends for my project (at least 1000 hrs. / year).
- d. I have very few family, job, community or household responsibilities.
- e. I don't have any extra cash to speak of, and will have to borrow the money.
- f. I have under $10,000 cash to invest and will not have to borrow this money.
- g. I have up to $50,000 to invest and will not have to borrow this money.
- h. I have a large support group of friends and professionals.
- i. I have discussed the possible hardships with my family & have their support.
- **Questions Totals:**

5. Can I handle the hardships, disappointments and rigors of Ironman Inventing?
- a. I am physically fit and spend more than 3 hours/week in aggressive physical activity.
- b. I am a positive, take no prisoners, strong ego, success-driven individual.
- c. I have had my ego bruised before - and recover fast; I can't be held down.
- d. I have competed in individual sports (swimming, cycling, running, skydiving, etc.)
- e. I have competed on a sports team (softball, basketball, rugby, etc.)
- f. I have planned my family, personal (away from the project), and invention time.
- g. I can be tough on people I think are taking advantage of me.
- **Questions Totals:**   **Total Combined Score: (add all lines) _____**

Q. 1: if all answers are in UNshaded boxes - you have your priorities right for success. Very few products reach commercial success just because the inventor *wanted* it to, or because a major medical cure was the sole objective - commit to building a profitable product business! Q.2.: if total score is < 15, give serious consideration to bringing in a partner with a score of >20. Q. 3.: if all your answers are in shaded boxes, you have the basic understanding of what your program may entail. Q.4.: c. If <3, rethink doing this program. Success will demand this amount of time.; (e, f, g) this is good to know, but if you do your plan thoroughly, you can attract needed capital - keep focused on your plan details.; (i) <5: watch out - inventing can destroy family and friendships. Q. 5.: < 25 - get on a training program now - NO JOKE. Statistics show how fitness is important to mental energy and physical endurance required in new business ventures!

PATENTCAFE™ PRODUCT / IDEAS ASSESSMENT FORM

This limited assessment is not meant to replace the complete Inveniton Assessment, and will not determine success potential.

General Product Description:

a.   My product / idea is: _____

b.   The purpose of this product is: _____

c.   My product is a q Consumer, q Industrial, q Educational Product, q Other: _____

d.   My product could be produced by a company with these S.I.C. codes (found in mailing list references): _____

Technical Product Analysis (indicate "Not True" with a justification explanation if a line item doesn't apply)

a.   My product is manufactured from; list all materials: _____

b.   My product incorporates new, exotic or novel materials; list all: _____

c.   The primary manufacturing processes used to produce my product are: _____

d.   Secondary manufacturing processes used to produce my product are: _____

e.   My product requires skilled professional, skilled technical or unskilled labor to produce; describe: _____

f.   My product will have to pass safety, environmental, transportation, electrical or other certification testing by a third
      party; specify: _____

g.   My product can be manufactured in a third-world country: _____

h.   My product has drawbacks; explain all technical problems you know of: _____

Competitive Product Analysis (indicate "Not True" with a justification explanation if a line item doesn't apply)

Listing of competitive products & manufacturers (include alternative products, ie: a stuffed toy "hero" doll needs to be compared to dolls other than stuffed [plastic molded, etc.] as well as dolls of "other heroes"; " Chain saw" needs to be compared against gas & electric chain saws as well as other "saws" [hand saws, Skilsaw® etc.]). List EVERY competitive manufacturer and competitive product on another sheet).

a.   My product has a lower production cost (& sales price) than competitive products because: _____

b.   My product provides more function or performance than the competition because: _____

c.   My product can be assembled faster then the competition's because: _____

d.   My product incorporates features of a second type of product; explain: _____

e.   My invention represents a "first of its type" product vs. an innovation; explain: _____

f.   My product is an improvement over the competition's (prior art) because; list all objective attributes: _____

g.   My product has competitive shortcomings; explain all known competitive deficiencies: _____

Proof of Concept: (indicate "Not True" with a justification explanation if a line item doesn't apply)

a.   I have a concept drawing of the product; (attach and include in your inventor's notebook); _____

b.   I have detailed engineering drawings or formula calculations of my product: _____

c.   I have built a working prototype of my product; take detailed photos: _____

e.   I have learned these production, material, cost, or other issues as a result of developing early concepts: include
      description of details or formulas rejected, and why, as a result of your early tries: _____

Download Patentcafe's Professional-level Invention Assessment from http://ww.ipbookstore.com.

SUPPLIER NON-DISCLOSURE AGREEMENT

Date _____

This agreement between _____ hereinafter referred to as INVENTOR and _____ hereinafter referred to as SUPPLIER is entered into under the following terms and conditions.

INVENTOR invites supplier to provide cost information for the following work: To perform engineering work on product concept as per sketches and specifications to be submitted. To enable SUPPLIER to perform this service, it is necessary for the INVENTOR to provide certain secret or confidential information (herein referred to as "Subject Matter") relating to his invention or product concept concerning a pet toy.

1. SUPPLIER agrees not to reveal, publish or communicate the Subject Matter to any other party for any purpose without the written consent of the INVENTOR.

2. SUPPLIER agrees to use this information strictly for the purpose of performing his service to the INVENTOR and agrees to hold it in the strictest confidence at all times.

3. All of the work done by the SUPPLIER in connection with the Subject Matter, whether or not patentable, is and shall remain the sole property of the INVENTOR

4. Upon completion of the assignment, if awarded, SUPPLIER agrees to return all material and objects that may have been provided by the INVENTOR, plus any copies he might have made.

5. If portions of the Subject Matter are already in the public domain, or if SUPPLIER can document that he has prior knowledge of the material from another source, he is not obligated to hold that specific material in confidence.

6. Except for possible exclusions indicated in Point 5, this Agreement shall be in force for five years, commencing with the above date. After this time, the obligations of confidentiality are cancelled.

This Agreement shall be construed in accordance with the laws of the state of Pennsylvania and contains the entire understanding of the parties hereto.

IN WITNESS WHEREOF, the parties have indicated their agreement to all of the above terms by signing and dating where below indicated.

INVENTOR: _____   SUPPLIER: _____
    Date: _____   Date: _____

Index